Introduction to Personal Computers for Networking Professionals

© 1998 · PC Age, Inc. All Rights Reserved · 20 Audrey Place · Fairfield, NJ 07004 · U.S.A. · Tel: 973-882-5002
www.pcage.com

Copyright © 1998 by PC Age, Inc. All rights reserved. No part of this work may be reproduced or transmitted in any form or by any means, electronic or mechanical, including photocopying or recording, or by any information storage or retrieval system without the prior written permission of PC Age, Inc., unless such copying is expressly permitted by federal copyright law. Address inquiries to PC Age, Inc., 20 Audrey Place, Fairfield, NJ 07004.

This book is sold as is, without warranty of any kind, either express or implied, respecting the contents of this book, including but not limited to implied warranties for the book's quality, performance, merchantability, or fitness for any particular purpose. Neither PC Age, Inc. nor its resellers shall be liable to the purchaser or any other person or entity with respect to any liability, loss, or damage caused or alleged to be caused directly or indirectly by this book. Further, PC Age, Inc. reserves the right to make changes to any and all parts of this book at any time, without obligation to notify any person or entity of such changes.

ISBN 1-57739-020-2

Table of Contents

WELCOME, PROGRAM OBJECTIVES

UNIT 1: MICROCOMPUTER CONCEPTS FOR NETWORK USERS

CHAPTER 1: INTRODUCTION TO COMPUTERS 1-1

- **WHAT A COMPUTER DOES** 1-1
  - INPUT OPERATION 1-1
  - PROCESSING 1-2
  - STORAGE 1-2
  - OUTPUT OPERATION 1-2
- **COMPUTER COMPONENTS** 1-2
  - INPUT DEVICES 1-3
  - SYSTEM UNIT 1-3
    - Microprocessor, or Central Processing Unit (CPU) 1-3
    - Primary Storage Memory 1-3
  - STORAGE UNITS 1-3
  - OUTPUT DEVICES 1-3
- **REPRESENTING PROGRAMS AND DATA IN MEMORY: THE BINARY SYSTEM** 1-4
  - ASCII (AMERICAN STANDARD CODE FOR INFORMATION INTERCHANGE) 1-5
  - EBCDIC (EXTENDED BINARY CODED DECIMAL INTERCHANGE CODE) 1-5
- **CHAPTER 1 REVIEW** 1-6

CHAPTER 2: INPUT DEVICES 2-1

- **KEYBOARD** 2-1
  - COMMON KEYBOARDS 2-1
    - XT Keyboard 2-1
    - AT Keyboard 2-1

  101-Key Keyboard 2-2
 THE KEYBOARD CONNECTOR 2-2
 MAINTENANCE 2-2
 TROUBLESHOOTING 2-2
MOUSE 2-3
 THE MOUSE CONNECTOR 2-3
 MAINTENANCE 2-3
 TROUBLESHOOTING 2-4
OTHER INPUT DEVICES 2-4
 MICROPHONES 2-4
 WIRELESS INPUT DEVICES 2-4

CHAPTER 3: MICROPROCESSOR 3-1

PARTS OF THE MICROPROCESSOR 3-1
 CONTROL UNIT 3-1
 ARITHMETIC/LOGIC UNIT (ALU) 3-2
MICROPROCESSOR PERFORMANCE 3-2
 SYSTEM CLOCK 3-2
 BUSES 3-3
 WORD SIZE 3-3
INTEL MICROPROCESSORS 3-3
 8086 3-4
 8088 3-4
 80286 3-4
  Virtual Memory 3-5
  Multitasking 3-5
  Real and Protected Mode 3-5
 80386 3-6
 80486 3-7
 PENTIUM P5 (80586) 3-7
 PENTIUM PRO 3-7
 PENTIUM II 3-8
 MMX TECHNOLOGY 3-8
MOTOROLA MICROPROCESSORS 3-8
 MC68000/MC68HC000 3-9
 MC68020 3-9
 MC68030 3-9
 MC68040 3-10
CHAPTER 3 REVIEW 3-11

CHAPTER 4: DATA BUS 4-1

    ISA (INDUSTRY STANDARD ARCHITECTURE) 4-1
    MCA (MICRO CHANNEL ARCHITECTURE) 4-2
    EISA (EXTENDED INDUSTRY STANDARD ARCHITECTURE) 4-2
    VESA (VIDEO ELECTRONIC STANDARDS ASSOCIATION) 4-3
    PCI (PERIPHERAL COMPONENT INTERCONNECT) 4-3
    COMPAQ FLEX ARCHITECTURE 4-4
    THE APPLE MACINTOSH BUS 4-4
  COMPUTER HARDWARE CONFIGURATION 4-5
    INTERRUPT REQUESTS (IRQ) 4-5
    DIRECT MEMORY ACCESS (DMA) 4-7
    I/O ADDRESS (INPUT/OUTPUT ADDRESS) 4-8
    BASE MEMORY ADDRESS 4-8
  CONFIGURING ELEMENTS 4-9
    DIP SWITCHES 4-10
    JUMPERS 4-10
    TERMINATING RESISTORS 4-10
    CMOS SETUP 4-11
    AUTOMATIC SETUP ROUTINE 4-11
    DEVICE DRIVERS 4-12
  CHAPTER 4 REVIEW 4-13

CHAPTER 5: COMPUTER MEMORY 5-1

  TYPES OF MEMORY 5-1
    RANDOM ACCESS MEMORY (RAM) 5-1
      Storing Data in RAM 5-2
      Dynamic RAM (DRAM) 5-2
      Extended Data Output RAM (EDO RAM) 5-2
      Static RAM (SRAM) 5-3
      Synchronous Dynamic RAM (SDRAM) 5-3
      Parity 5-3
      Fake Parity 5-4
      Error Correction Code (ECC) 5-5
    READ-ONLY MEMORY (ROM) 5-5
    PROGRAMMABLE READ-ONLY MEMORY (PROM) 5-6
    ERASABLE PROGRAMMABLE READ-ONLY MEMORY (EPROM) 5-6
    ELECTRICALLY ERASABLE PROGRAMMABLE READ-ONLY MEMORY
    (EEPROM) 5-6

SYSTEM MEMORY 5-6
   CONVENTIONAL MEMORY 5-7
   EXTENDED MEMORY 5-8
   EXPANDED MEMORY 5-9
   UPPER MEMORY 5-10
   HIGH MEMORY AREA (HMA) 5-11
   DIRECT MEMORY ACCESS (DMA) 5-11
CHAPTER 5 REVIEW 5-12

CHAPTER 6: AUXILIARY MEMORY—STORAGE DEVICES 6-1

DISKS 6-1
   FLOPPY DISKS 6-1
      Disk Capacity and Density 6-2
   HARD DISKS 6-3
      Partitioning the Hard Disk 6-4
      Random Access Time 6-5
      Hard Disk Controllers 6-5
         ST506/412 6-6
         Enhanced Small Device Interface (ESDI) 6-6
         Small Computer System Interface (SCSI) 6-7
         Integrated Drive Electronics (IDE) Interface 6-7
OTHER STORAGE DEVICES 6-7
   CD-ROM 6-7
   TAPE BACKUP 6-8
      Backup Strategies 6-8
         Full Backup 6-9
         Incremental Backup 6-9
         Differential Backup 6-9
      Restore Strategies 6-10
CHAPTER 6 REVIEW 6-11

CHAPTER 7: VIDEO DISPLAY UNIT 7-1

VDU FEATURES 7-1
   DISPLAY QUALITY 7-1
   PIXELS AND RESOLUTION 7-1
   COLOR 7-2
MONITOR TYPES AND DISPLAYS 7-2
   MONOCHROME MONITORS 7-3
      TTL (Transistor-Transistor Logic) 7-3
      Composite Monochrome 7-3

VGA Monochrome 7-3
Multiscanning Monochrome 7-3
COLOR MONITORS 7-4
RGB 7-4
CGA (Color Graphics Adapter) 7-4
EGA (Enhanced Graphics Adapter) 7-4
VGA (Video Graphics Array) 7-4
Multiscanning Color Displays 7-5
CHAPTER 7 REVIEW 7-6

CHAPTER 8: INPUT/OUTPUT PORTS 8-1

PARALLEL PORTS 8-1
SERIAL PORTS 8-2
SERIAL DEVICE TYPES 8-2
COMMUNICATION SYNCHRONIZATION METHODS 8-3
Asynchronous Communication 8-3
Synchronous Communication 8-3
BPS AND BAUD RATE 8-4
HANDSHAKING 8-5
FLOW CONTROL 8-5
CHAPTER 8 REVIEW 8-6

UNIT 2: DOS FOR NETWORK USERS

CHAPTER 9: DISK OPERATING SYSTEM (DOS) 9-1

COMPONENTS OF DOS 9-1
DOS CORE OPERATING SYSTEM 9-1
IBMBIO.COM 9-2
IBMDOS.COM 9-2
COMMAND.COM 9-2
AUXILIARY UTILITY PROGRAMS 9-2
THE COMPUTER STARTUP PROCESS (BOOT) 9-2
COLD BOOT PROCESS 9-3
WARM BOOT PROCESS 9-3
POWERING ON YOUR COMPUTER 9-4
DOS SEARCH 9-4
DOS CONFIGURATION 9-4

© 1998 · PC Age, Inc. All Rights Reserved · 20 Audrey Place · Fairfield, NJ 07004 · U.S.A. · Tel: 973-882-5002
www.pcage.com

CHAPTER 10: DOS INTERNAL COMMANDS 10-1

USING DOS COMMANDS 10-1
   F1 AND F3 FUNCTION KEYS 10-1
HELP AND FASTHELP 10-1
COMMAND PARAMETERS 10-2
   SYNTAX 10-2
THE DOS COMMAND PROCESSOR (COMMAND.COM) 10-3
INTERNAL DOS COMMANDS 10-3
   CLS 10-4
   VOL 10-4
   VER 10-4
   DIR 10-4
   DATE AND TIME 10-6
   PROMPT 10-6
   MD (MAKE DIRECTORY) 10-8
   CD (CHANGE DIRECTORY) 10-8
   RD (REMOVE DIRECTORY) 10-9
   COPY 10-9
   REN (RENAME) 10-10
   DEL (DELETE) AND ERASE 10-10
   TYPE 10-10
   PATH 10-12

CHAPTER 11: DOS EXTERNAL COMMANDS 11-1

ATTRIB 11-1
LABEL 11-2
EDIT 11-3
MOVE 11-3
DISKCOPY 11-4
UNDELETE 11-4
SYS 11-5
FORMAT 11-5
   FORMATTING FLOPPY DISKETTES 11-5
   FORMATTING HARD DRIVES 11-6
UNFORMAT 11-6
XCOPY 11-7
TREE 11-8
DELTREE 11-8
MEM 11-8
CHKDSK 11-9

SCANDISK	11-9
MSD	11-9
PRINT	11-10
SHARE	11-10
MSBACKUP	11-10
FDISK	11-11
OTHER DOS CONVENTIONS	11-11
|MORE	11-11
DOSKEY	11-11

CHAPTER 12: DOS DIRECTORIES AND FILES 12-1

DIRECTORY	12-1
SUBDIRECTORY	12-1
FILES	12-2
SPECIAL DOS CHARACTERS	12-3
FILE NAME EXTENSIONS	12-3
WILDCARD CHARACTERS	12-4
ASTERISK (*)	12-5
QUESTION MARK (?)	12-5
BATCH FILES	12-6
CREATING A BATCH FILE	12-6
REM	12-7
ECHO	12-7
PAUSE	12-8
DOS CONFIGURATION FILES	12-8
THE CONFIG.SYS FILE	12-8
THE AUTOEXEC.BAT FILE	12-8

CHAPTER 13: DISKS AND DRIVES 13-1

HARD DISKS AND FLOPPY DISKS	13-1
FLOPPY DISKS	13-1
Disk Capacity and Density	13-2
HARD DISKS	13-2
Partitioning the Hard Disk	13-3
Formatting the Hard Disk	13-4

CHAPTER 14: MEMORY MANAGEMENT 14-1

MEMORY OPTIMIZATION 14-1
F5 AND F8 FUNCTION KEYS 14-1
CONFIG.SYS 14-2
BREAK 14-3
BUFFERS 14-4
DEVICE 14-5
DEVICEHIGH 14-6
DOS 14-7
FCBS 14-7
FILES 14-8
INSTALL 14-9
LASTDRIVE 14-9
REM 14-10
SHELL 14-10
STACKS 14-11
Sample CONFIG.SYS File 14-12
AUTOEXEC.BAT 14-12
@ 14-13
CALL 14-13
ECHO 14-13
PAUSE 14-14
PATH 14-14
PROMPT 14-15
REM 14-15
MEMORY OPTIMIZATION TIP 14-15
MSCDEX.EXE 14-15
MULTIPLE CONFIGURATION 14-16
MENU BLOCKS 14-17
MENUITEM 14-18
MENUCOLOR 14-18
MENUDEFAULT 14-19
Configuration Blocks 14-19

CHAPTER 15: EMERGENCY PREPAREDNESS 15-1

BOOTUP ORDER/PROCEDURE 15-1
BIOS SETUP 15-1
UNIT 2 REVIEW 15-4

UNIT 3: WINDOWS

CHAPTER 16: WINDOWS 95 .... **16-1**

    NEW FEATURES 16-2
    INSTALLING WINDOWS 95 16-2
    RUNNING WINDOWS 95 16-3
    USING WINDOWS 95 16-4
        RUNNING PROGRAMS 16-5
            Accessories 16-6
            Startup 16-6
            MS-DOS Prompt 16-7
            Windows Explorer 16-7
    SHUTTING DOWN YOUR COMPUTER 16-7
    INSTALLING SOFTWARE 16-8
        INSTALLING SOFTWARE USING THE RUN COMMAND 16-8
        INSTALLING SOFTWARE USING ADD/REMOVE PROGRAMS 16-9
    INSTALLING HARDWARE 16-11
        ADDING PRINTERS 16-12
    CREATING SHORTCUTS 16-14
        CREATING SHORTCUTS USING WINDOWS EXPLORER 16-14
        CREATING SHORTCUTS USING THE FILES OR FOLDERS COMMAND 16-15
        CREATING SHORTCUTS USING THE RIGHT MOUSE BUTTON 16-16
    SETTING UP A NETWORK 16-16
        CLIENT SOFTWARE 16-17
        NETWORK ADAPTER 16-17
        PROTOCOL 16-18
        SERVICES 16-18
    SHARING FOLDERS AND PRINTERS 16-20
        SHARING A FOLDER IN SHARE-LEVEL ACCESS CONTROL 16-21
        SHARING A PRINTER IN SHARE-LEVEL ACCESS CONTROL 16-21
    MAINTAINING AND OPTIMIZING YOUR SYSTEM 16-21
        SCANDISK 16-22
        DISK DEFRAGMENTER 16-23
        DRIVESPACE 16-24

APPENDIX A: MICROPROCESSOR COMPARISON TABLES **A-1**

APPENDIX B: DOS ERROR MESSAGES **B-1**

Welcome!

We are pleased to welcome you to PC Age's Introduction to Personal Computers for Networking Professionals training program.

The length of this program is four days (full time) or eight evenings (part time).

All students attending this program should have already viewed all three DOS videos.

This course provides a global overview of the personal computer, including the system hardware most typically used, the commonly used software, and two of Microsoft's operating systems: Microsoft DOS 6.22 and Microsoft Windows 95.

Through instructor lectures and student exercises with an emphasis on hands-on application, you have an opportunity to explore the physical components and various operating systems of the modern PC.

If you have any problems or concerns during any portion of this training program, please do not hesitate to inform your instructor.

We will begin our training program by first reviewing the Program Objectives and the Course Outline.

Thank you for attending!

PC Age

Program Objectives

In this course you will learn:

- The standard components of a personal computer
- The functions of the major components of a personal computer
- The important performance-related PC computer features
- How to install DOS
- The basic features of Microsoft DOS
- How to check and set your system's date and time
- How to check the version of your current operating system
- How to delete or rename files
- How to delete or rename directories
- How to copy your files from one directory to another or from one drive to another
- How to copy the complete directory structure, including all files and subdirectories
- How to delete the complete directory structure, including all files and subdirectories
- How to format a floppy disk and make it a boot disk
- How to format a disk and transfer the system simultaneously
- How to use wildcard characters to ease your different tasks
- How to use function keys to minimize your keystrokes
- How to get help for DOS commands
- How to install and use Windows 95
- The basic features of Windows 95
- How to use and work with Windows

- The functions of the default groups in Windows 95
- How to install applications in Windows 95
- How to open applications using Windows

Unit 1: Microcomputer Concepts for Network Users

Chapter 1   Introduction to Computers

A computer is an electronic device, operating under the control of instructions stored in its memory unit, that can accept data, process data, produce output, and store the results.

The word "data" refers to the numbers and words given to the computer during the input operation. The manipulation of data into a usable form is called data processing.

What a Computer Does

The functions of any computer can be classified as one of four basic operations:

- Input Operation
- Processing
- Storage
- Output Operation

Input Operation

Input is the data the user enters into the computer; input operation is the computer's acceptance of this data. Input devices include keyboard, mouse, light pen, scanner, voice-recognition system, etc.

Processing

When a computer produces information by performing various operations on data, it is said to be processing the data. The operations performed on data include sorting, calculating, comparing, and recording.

Storage

Computers use storage devices to permanently save information in an electronic format for future use. Devices used for storage include floppy disk, hard disk, magnetic tape, CD-ROM, DVD, removable media, etc.

Output Operation

Output refers to converting the information that is located in the computer into a form that people can use. For example, printed reports and screen displays are considered output. Output devices include printer, monitor, plotter, etc.

Computer Components

The components of a computer are the physical parts which computer uses to process data. The basic computer hardware components are:

- Input devices
- System unit
- Storage unit
- Output devices

Input Devices

Input devices include any equipment used to enter data into a computer—keyboard, mouse, light pen, scanner, etc.

System Unit

The system unit contains the electronic circuits that cause the processing of data to occur. The system unit is made up of two basic parts.

Microprocessor, or Central Processing Unit (CPU)

The CPU performs arithmetic operations (such as addition, subtraction, multiplication, and division) and logical operations (such as less than, greater than, and equal to).

Primary Storage Memory

The primary storage memory stores data and program instructions electronically while they are being processed. by THE CPU

Storage Units

The storage units (also called auxiliary storage) store data and programs when they are not being used by the processor unit. Storage units include floppy disks, hard disks, and magnetic tapes.

Output Devices

Output from a computer can be presented in many forms. The most commonly used output devices are printers, monitors, and plotters.

Representing Programs and Data in Memory: The Binary System

Program instructions and data are stored in computer memory as binary numbers. These numbers are called bits (short for BInary digiTs). The binary number system (base 2) represents quantities by using only two symbols, 0 and 1. So a bit can represent only one of two values—either 0 or 1, which are used to signify OFF and ON, respectively.

In most microprocessors, binary numbers are grouped into eight-bit units called bytes (short for BinarY digiT Eights). If a particular byte is part of a program instruction, then the number is an operation (such as ADD or MOVE) followed by another number representing a location in memory. If a byte is a data item, then the number usually represents a character (such as a letter of the alphabet, a number from 0 to 9, or a punctuation mark).

Generally speaking, when we consider characters stored in main memory, we think of each character being stored in one memory location, or byte. Each alphabetic, numeric, and special character stored in the memory of the computer is represented by a combination of OFF and ON bits. The computer can distinguish between characters because the combination of OFF and ON bits assigned to each character is unique. Two popular codes that use combinations of zeros and ones for representing characters in memory are the ASCII code and the EBCDIC code. Rather than showing a single unbroken string of ones and zeros, these codes are usually shown as groups of four bits, to use hexadecimal form.

ASCII (American Standard Code for Information Interchange)

ASCII is the coding system most widely used to represent data. This was the bit pattern code selected by IBM for use in its first PC. Regular ASCII code uses only seven bits to represent 128 characters; the eighth bit is used for error checking. The IBM extended character set uses the eighth bit to define an additional 128 characters (for a total of 256 characters). Most microcomputer manufacturers now support the extended character set.

EBCDIC (Extended Binary Coded Decimal Interchange Code)

As opposed to the ASCII code, which is widely used on personal computers and many minicomputers, EBCDIC is primarily used on IBM mainframes. The EBCDIC code uses eight bits to represent a character.

Chapter 1 Review

Q1. What are the four basic operations of a computer?
Input   output   Storage + Processing

Q2. What are the four basic components of a computer?
Input Devices,   System Unit   Storage Unit
Output Devices

Q3. How are instructions and data stored in a computer?
As Binary Numbers

Q4. How are characters represented in a computer?
bytes (Binary Digit Eights

Chapter 2   Input Devices

Keyboard

The keyboard is still the main input device used to enter data into the computer. Its configuration derives from the typewriter, which uses a key arrangement very similar to the one on today's keyboards. The PC's keyboard contains a microprocessor chip and circuitry. It is susceptible to mechanical damage, primarily from environmental factors.

Common Keyboards

On IBM compatible computers there are three basic types of keyboards, as well as some newer variations.

XT Keyboard

The original PC or IBM XT had an 83-key keyboard, with the numeric keypad and the cursor control keys using the same keys on the right side of the keyboard.

AT Keyboard

This keyboard improved on the XT design with a larger <ENTER> key.

101-Key Keyboard

Referred to as the "enhanced keyboard", this keyboard design separated the cursor and numeric keys and added two function keys. This design has generally superseded the earlier keyboard types.

Newer keyboard designs—including the Microsoft Natural Keyboard, which features a "broken" or separated keyboard layout—have been developed to improve ergonomics and reduce wrist strain.

The Keyboard Connector

There is usually a cable connecting the keyboard to the system unit. The keyboard cable is plugged into the computer with a round five-pin plug.

Maintenance

The major maintenance procedure for keyboards is to avoid spilling anything into them. Occasionally wiping the keyboard and blowing dust off of it aids in preventive maintenance.

Troubleshooting

If you are having problems with your keyboard, there are several things you can check.

Is the keyboard plugged securely into the correct port? The mouse port sometimes looks identical, and the keyboard can easily be plugged into it by accident. This will not only prevent the keyboard from working, it may damage the keyboard.

Is there a problem with all of the keys, or with just one key? If only one key is malfunctioning, the problem could be caused by a stuck spring or dirty contact. The key can be removed by pulling it up. Check the springs and contacts, then put the key back by snapping it into place.

If the problem is with all of the keys, try replacing the keyboard. Keyboards are not expensive, so replacing them is often more efficient than fixing them.

Mouse

The mouse has become ubiquitous in Windows-based computers, although there is a growing trend for keyboards—especially in notebook computers—to incorporate pointing devices. Most mice sense mechanically when they are being moved, and transmit that movement to the computer. In addition, there are typically two or more buttons used in pointing, selecting, drawing, etc. As you move the mouse, the cursor on the screen follows the mouse's movements.

The Mouse Connector

There is usually a cable connecting the mouse to the system unit. The mouse cable is plugged into the computer with one of two kinds of plugs: a round five-pin plug similar or identical to the one for the keyboard, or a serial plug that connects to the serial port.

Maintenance

Like keyboards, the major maintenance procedure for mice is to keep them clean. The surface on which a mouse is used must also be kept very clean. Keep mice away from the debris left by pencil erasers, and from anything else that could get the mouse and its trackball sticky.

Troubleshooting

In a mechanical mouse, the trackball and the wheels turned by the trackball sometimes get jammed by dirt. Remove the ball and clean the mechanism with rubbing alcohol and a cotton swab.

Other Input Devices

Microphones

The microphone has now become an input device, using software that translates the spoken word into text, numbers, and commands.

Wireless Input Devices

A wireless input device can provide a cordless interface with a computer. Most wireless input devices use infrared signals to transmit to a receiver connected to the system unit.

Chapter 3   Microprocessor

The microprocessor can be thought of as the "brain" of the computer. Physically, a microprocessor is a silicon chip composed of integrated circuits and transistors. The microprocessor is also called the Central Processing Unit (CPU). Minicomputers and mainframes use additional chips for their CPUs.

Parts of the Microprocessor

There are two basic parts of the microprocessor: the control unit and the arithmetic/logic unit.

Control Unit

Just as the human brain controls the body, the control unit controls the computer. The control unit operates by repeating the following operations:

- Fetching – obtaining the next program instruction from main memory
- Decoding – translating the program instruction into commands the computer can process
- Executing – processing the command
- Storing – writing the result of the command to main memory

Faster, newer processors have further divided the number of steps in each operation, thereby allowing more sequential operations at the same time. The Pentium Pro uses 14 execution steps, and one step does not have to be completed before the next is begun.

Arithmetic/Logic Unit (ALU)

This unit contains the electronic circuitry necessary to perform arithmetical and logical operations on data. Arithmetical operations include addition, subtraction, multiplication, and division. Logical operations consist of comparing one data item to another to determine if the first data item is greater than, equal to, or less than the other.

Microprocessor Performance

The speed at which a computer can execute its functions is influenced by three factors:

- System clock
- Buses
- Word size

System Clock

The control unit utilizes the system clock to synchronize (regulate the timing of) all computer operations. The system clock generates electronic pulses at a fixed rate, measured in MegaHertz (MHz). One MegaHertz equals one million pulses, or cycles, per second. The speed of the system clock varies among computers. Some personal computers can operate at speeds in excess of 400 MegaHertz. The processor chip in most newer computers (many 486s, all Pentiums, and equivalents of both) run at 1.5 or more times the speed of the rest of the system.

Buses

Any path along which bits are transmitted is called a bus (or a data bus). Buses usually transfer 8, 16, 32, or 64 bits at a time. An eight-bit bus has eight lines and can transmit eight bits at a time. On a 16-bit bus, bits can be moved from place to place 16 bits at a time; on a 32-bit bus, bits are moved 32 bits at a time. The larger the number of bits that are handled by a bus, the faster the computer can transfer data. Some newer PCs have a 128-bit data bus.

Word Size

The word size is the number of bits the computer can process at one time. Processors can have word sizes of 8, 16, 32, or 64 bits. A processor with an eight-bit word size can manipulate eight bits at a time. For example, the ALU of an eight-bit processor requires four operations to add two numbers of four digits each, because a separate operation is required to add each of the four digits. A 16-bit processor would require only two operations, and a 32-bit processor would require only one.

Intel Microprocessors

IBM and IBM compatible computers are built around the following microprocessors, manufactured by the Intel Corporation:

- 8086
- 8088
- 80286
- 80386

- 80486
- Pentium P5 (80586)
- Pentium Pro
- Pentium II

8086

The 8086 was introduced in 1978. It has the following features:

- 16-bit microprocessor (16-bit word size)
- 16-bit data bus
- 20-bit memory address that allows the 8086 direct control of one megabyte of memory
- Clock speed of 8 MHz

8088

The 8088 was introduced in 1979. It was basically the same as the 8086, except that it was designed for an 8-bit data bus. The 8088 was used in the original IBM PC. It has the following features:

- 16-bit microprocessor (16-bit word size)
- 8-bit data bus
- 20-bit memory address
- Clock speed of 4.77 to 10 MHz

80286

The 80286 was introduced in 1984. It has the following features:

- 16-bit microprocessor (16-bit word size)
- 16-bit data bus
- Ability to access up to 16 MB of real memory
- Ability to simulate up to 1 GB of virtual memory

- Multitasking
- Real and protected operating modes
- Clock speed of 6 to 20 MHz

The 80286 introduced three new features that are associated with it and with all higher Intel microprocessors: virtual memory, multitasking, and real and protected modes.

Virtual Memory

Virtual memory increases the effective limits of memory by expanding the amount of main memory to include disk space. Without virtual memory, an entire program must be loaded into main memory during execution. With virtual memory, only the portion of the program that is currently being used is required to be in main memory.

Multitasking

Multitasking allows more than one program to be run at the same time. Even though the CPU is only able to work on one program instruction at a time, its ability to switch back and forth between programs makes it appear that all programs are running at the same time. For example, a computer with a multitasking operating system could be performing a complex spreadsheet calculation and downloading a file from another computer at the same time that the user is writing a memo with a word processing program.

Real and Protected Mode

The 80286 and 80386 processors can run in either of two modes: real or protected. In real mode, the processor emulates an 8086/8088 processor and runs as though it actually were an 8086/8088 processor. When an 80286 or 80386 processor runs in real mode, it can access a maximum of 1 MB of RAM and can run only one software application at a time.

In protected mode, 80286 and 80386 processors are not subject to the same memory constraints as an 8086 processor. The 80286 uses a 24-bit address bus. Raising the base number of 2 to the power of 24 (because a bit can have one of two values) yields 16,777,216. This is the number of unique memory addresses in the computer, each of which can store one byte of information; so an 80286 can access up to 16 MB of real memory (16,777,216 bytes equals 16 MB). The 80386 processor uses a 32-bit address bus. Raising the base number of 2 to the power of 32 yields 4,294,967,296 unique memory addresses, each of which can store one byte of information, so the 80386 can access up to 4 GB of real memory.

Protected mode operation also provides the capability of multitasking.

80386

The 386 chip has the following features:

- 32-bit microprocessor (32-bit word size)
- Ability to access up to 4 GB of real memory
- Ability to simulate up to 64 TB (terabytes) of virtual memory
- Multitasking
- Real and protected operating modes
- Clock speed of 16 to 32 MHz

The 80386 introduced 32-bit processing. When this chip was introduced, most PC expansion boards were still 8- or 16- bit and were unable to use the 80386's full capabilities. As a result, Intel introduced another version of the 80386 called the 80386SX. The original version was called the 80386DX. The SX chip has a 16-

bit external interface and is limited to the 16-bit bus architecture, whereas the DX chip has a 32-bit external interface and can be used on a modified AT bus to accommodate the 32-bit chip.

80486

The 80486 has the following features:

- 32-bit microprocessor (32-bit word size)
- 32-bit data bus
- Math coprocessor included in the microprocessor
- Special high-speed memory in the cache controller
- Clock speed of up to 64 MHz

Intel introduced two kinds of 486: the 80486SX and the 80486DX. The main difference between these two microprocessors is that the 80486DX has a built-in math coprocessor.

Pentium P5 (80586)

Intel introduced the Pentium P5 or 80586 chip in 1993, using the name Pentium instead of 80586 to distinguish it from other clone manufacturers. The Pentium includes two 80486-type CPUs for true dual processing, processing speeds of over 200 MHz, a 64-bit wide internal interface, and independent access to multiple internal data caches. It uses a 32-bit data bus.

Pentium Pro

Intel introduced the Pentium Pro chip as the successor to the Pentium. It is actually two chips: a main processor and a built-in 256 KB external cache (L2 cache). The Pentium Pro features a 64-bit wide internal interface, and independent access to multiple internal data caches. It uses a 32-bit data bus. Processing speeds are up to 300 MHz.

Pentium II

Intel's Pentium II processor delivers excellent performance for all PC software, and features speeds of up to 400 MHz. It incorporates full multimedia capabilities including enhanced color, realistic graphics, and full-screen, full-motion video. It has a 32 K (16 K/16 K) L1 cache to provide fast access to heavily used data, and a 512 K L2 cache improves performance by reducing the average memory access time and providing fast access to recently used instructions and data.

MMX Technology

Intel has incorporated additional processor code on MMX (Pentium and Pentium Pro) chips to speed the operation of multimedia and communications applications. This technology is compatible with Windows 3.1, Windows 95, and Windows NT.

Note: All later Intel chips are compatible with previous Intel chips.

Motorola Microprocessors

The Apple Macintosh uses microprocessors built by Motorola. This family of microprocessors includes the following:

- MC68000
- MC68HC000
- MC68020
- MC68030
- MC68040

MC68000/MC68HC000

These two microprocessors were introduced in 1979. Their features include:

- 16-bit data bus
- 24-bit internal address bus
- MB direct memory access
- Clock speed of 8 to 16 MHz

MC68000 is used in Macintosh 128K, 512K, 512Ke, Plus, SE, and Mac Classic. Mac portable uses MC68HC000, which is a low-power version of MC68000. MC68000 was also used in the first Novell NetWare file server.

MC68020

This microprocessor is fully compatible with earlier M68000 processors. Its features include:

- 32-bit data bus
- 32-bit internal address bus
- 4 GB direct memory addressing
- Clock speeds of 12 and 16 MHz
- Math coprocessor support

This microprocessor is used in LC and Macintosh II computers.

MC68030

MC68030 offers twice the performance of MC68020. Its features include:

- Two independent 32-bit address buses
- Two independent 32-bit data buses

- Data cache and instruction cache on chip
- Allows the manufacturing of dual-processor computers
- Ability of a single processor to perform two processes simultaneously
- Clock speed of 16 to 40 MHz

This microprocessor is used in SE/30, IIx, IIcx, IIci, IIsi, and IIfx computers.

MC68040

MC68040 is the latest Motorola microprocessor. It offers 64-bit internal interface, 64-bit data bus, and clock speed of 64 MHz.

Chapter 3 Review

Q1. What are the subunits of a microprocessor?

Control Unit + Arithmetic Logic Unit (ALU)

Q2. How does the bus size affect the processing speed?

The more bits that are handled by a bus the faster the data transfer is.

Q3. Compare the Intel 80286 and the Motorola 68000 microprocessors.

Both have 16 bit data bus, clock speeds are close. Both have 24 bit address bus (16 mB). Real memory. Very similar microprocessors.

Q4. What is the difference between real mode and protected mode?

Real - 286 or 386 processor emulates 8086 processor. No multitasking + only 1MB RAM access.
Protected: Allows multitasking + full use of 24 (286) or 32 (386) bit address bus.

Chapter 4   Data Bus

The data bus provides a path for data transfer between the CPU and other system components (memory, disks, video, and peripherals). In fact, the data bus connects all system components so they can communicate.

One of the most important components of the data bus architecture is the expansion bus or expansion slots. These expansion slots communicate directly with the memory and the microprocessor. External devices—including disk controller, video card, modem, printer, additional memory card, and network interface card—are connected to the system through expansion slots.

There are four main data bus standards:

- ISA
- MCA
- EISA
- VESA

ISA (Industry Standard Architecture)

IBM introduced the IBM PC/AT (Advanced Technology) in 1984. It was tied to the Intel 80286 microprocessor. The AT expansion bus is considered the industry standard. It is commonly known as the ISA bus. ISA offers an 8-bit or 16-bit data/expansion path. Most 286, 386, and 486 computers and their clones use ISA. 16-bit ISA does not take advantage of the 32-bit processing capabilities of the 386 microprocessor, but it is a good choice for workstations and file servers with a light network load. ISA is backward compatible with the 8-bit IBM XT bus, so IBM XT 8-bit boards can be used with an ISA expansion bus.

MCA (Micro Channel Architecture)

IBM introduced MCA in 1987, along with a new line of personal computers called PS/2, to provide an adequate expansion bus to work with 80386 and later microprocessors. MCA offers many advantages, including 32-bit throughput and the ability to automatically reconfigure the system when expansion boards are added. But it is not compatible with the original PC and AT bus designs, so PC and AT expansion boards will not work in an MCA machine.

Most IBM Personal System 2 (PS/2) computers with chips of 386 or later are built around Micro Channel Architecture. But the models 25 and 30 machines are built around the ISA bus. The models 50 and 60 use a 16-bit version of the MCA bus, while the models 70, 80, and 90/95 use a 32-bit version.

EISA (Extended Industry Standard Architecture)

The EISA bus was introduced in response to IBM's proprietary 32-bit MCA. It provides 32-bit throughput by expanding the ISA bus. EISA offers 32-bit throughput and auto-configuration, and also maintains backward compatibility with ISA bus. This means that in an EISA computer, you can either use 32-bit boards to take advantage of 32-bit architecture, or you can use 16-bit ISA boards, which are not supported by MCA.

Prior to the wide availability of PCI buses, computers that used an EISA bus were considered the ideal for busy file servers. A 32-bit network interface card and a 32-bit disk controller were used to take advantage of the 32-bit throughput. The organization that developed EISA (known as the Group of Nine) includes Compaq, AST, Epson, HP, NEC, Olivetti, Tandy, Wyse, and Zenith. The major problem with the EISA bus is that it is not a local bus, and operates at only 8 MHz. Thus, EISA tends to appear in machines in combination with a local bus type, either VLB or PCI (see below).

VESA (Video Electronic Standards Association)

The VESA bus standard is named after the industry group that promoted this local bus standard: the Video Electronic Standards Association. Also referred to as the VLB (VESA Local Bus), this standard provides a dedicated 32-bit line directly from the VESA slot to the microprocessor. This technology is called "local bus". VESA Local Bus throughput is much better than either EISA or ISA. However, VESA Local Bus throughput is a 32-bit, high-speed version of a limited ISA architecture. All data transfer must be through the CPU, keeping it in the middle of the system and thereby creating a bottleneck to system performance. VLB systems also require jumper settings.

PCI (Peripheral Component Interconnect)

In order to accommodate the growing data transfer demands of faster processors, expanded RAM, high resolution video, audio applications, high speed LAN interfaces, SCSI host adapters, and other peripherals, Intel designed the PCI (Peripheral Component Interconnect) bus. Features of the PCI bus include:

- A 64-bit wide data path, which also supports 32-bit data paths.
- Higher speed (faster transfer rates)

- Support for multiprocessor systems
- Support for bus-mastering, reducing the need to use CPU time for data transfers
- Support for the Plug and Play standard, enabling setup through software rather than through jumpers and DIP switches
- Ability to be combined with other bus slots on the same motherboard to make up for the fact that PCI slots do not accommodate other types of parts

There are several other bus standards, including Compaq's Flex Architecture and Apple's NuBus.

Compaq Flex Architecture

Compaq Computer Corporation introduced flex architecture to provide a faster means of moving data between the microprocessor and the memory. It used a separate data path that ran directly from the microprocessor to the memory, bypassing the expansion bus entirely. Flex design was based on a solution called DMA (Direct Memory Access).

The Apple Macintosh Bus

The Macintosh data bus differs dramatically from the bus architectures discussed above. It is called NuBus architecture, and was originally designed by Texas Instruments Inc.

Most Macintosh II computers are built around the NuBus design. However, the Mac SE does not use NuBus. Instead it uses a proprietary processor slot called the Processor Direct Slot (PDS). The PDS gives direct access to the processor from one expansion card. This gives a data transfer rate much faster than that provided by NuBus slots. The disadvantage of PDS architecture is that it supports only one PDS card, and that card must be processor-specific. This means that a Motorola 68030 PDS card will not work with a Motorola 68040 computer.

The Mac IIfx and some of the Quadra computers use six NuBus slots and one Processor Direct Slot.

Computer Hardware Configuration

As we can see, a microcomputer is composed of many components (or devices). The CPU and the other components continuously communicate in order to transfer and perform operations on data. With so many devices operating at once, there is a chance that devices will conflict with each other when communicating with the CPU. There are four parameters that can result in a conflict:

- IRQ
- DMA
- I/O address
- Base memory address

Interrupt Requests (IRQ)

An interrupt request is a control signal used by a device to let the CPU know that the device is requesting a service or has completed an operation. When the CPU receives an interrupt signal, it stops what it is doing and responds to the device which sent the signal, then returns to what it was doing before it received the interrupt. Each device must have a unique IRQ assigned so that the CPU knows which device sent the signal. For this reason, when adding devices on your computer, it is important to configure them with interrupts that have not already been assigned. A common PC problem is an interrupt conflict where two devices are trying to use the same IRQ; in this case, one or both devices will fail to work. Computers usually have 16 IRQs available.

When dealing with IRQs, another important concept to remember is "cascading". Computers use two banks of eight IRQs each. IRQ 2 (on the first bank) links the first bank to the second. If IRQ 2 is used for this purpose, it cannot be used by devices. Because some devices can use IRQs only from the first bank, it is not a good idea to use an IRQ from the first bank as a gateway to the second bank. That is why IRQ 9 is addressed as IRQ 2. This relationship is called cascading. IRQ 9 cannot be used by devices if IRQ 2 is used.

The following is a list of the commonly used IRQ assignments in a typical AT-compatible PC.

IRQ	Assignment
0	System timer
1	Keyboard
2	Used in AT-type PCs as a gateway to IRQs 9 to 15; some EGA/VGA video and ARCnet cards use this IRQ.
3	Serial port COM2 and COM4, often used for a Network Interface Card or a bus mouse
4	Serial port COM1 and COM3, usually used for a modem
5	Parallel printer port LPT2, sometimes used for a Network Interface Card or a sound card
6	Floppy drive controller
7	Parallel printer port LPT1, usually used by a printer
8	System clock

9	Redirected as IRQ 2
10	Available, if not used for a primary SCSI controller
11	Available, if not used for a secondary SCSI controller
12	Available, sometimes used for a PS/2 mouse
13	Math coprocessor
14	Hard drive controller
15	Available, often used for a second hard drive controller

Direct Memory Access (DMA)

Some devices can write data into the PC memory directly, without CPU intervention. This is called Direct Memory Access (DMA). DMA increases performance.

The old XT-type computers had four DMA channels (0-3) while the AT-type computers have eight channels (0-7). Some of these channels are used by standard PC components (hard disk, floppy disk controllers). The rest can be used by other devices. DMA channels can be shared by devices that do not need to use them simultaneously.

The following is a list of DMA channels and their uses.

DMA Channel	Assignment
0	Dynamic memory refresh
1	Hard disk controller
2	Floppy disk controller
3	Available
4-7	Available on AT and PS/2 computers

I/O Address (Input/Output Address)

Each device needs a unique I/O address (also called Base I/O or I/O Port) to communicate with the CPU. An I/O address is a reserved address in memory. I/O addresses are like post office boxes. If a device has data for the CPU, it puts the data in an I/O address, just as mail is put into a post office box. When the CPU wants to read data from that device, it looks in the I/O address, like a person taking mail out of a post office box. CPUs know devices through their I/O addresses. The communication between the CPU and the device is actually communication between the CPU and the I/O address of that device.

The following is a list of the commonly used I/O address ranges.

I/O Address	Assignment
000 - 00F	Reserved for the system
1F0 - 1F8	Hard disk controller
200 - 207	Game port
278 - 27F	LPT2
2F8 - 2FF	COM2
378 - 37F	LPT1
3F0 - 3F7	Floppy disk controller
3F8 - 3FF	COM1

Base Memory Address

Some controller cards and LAN boards need a reserved area of PC memory for loading on-board ROM instructions. Some LAN boards may also have a little RAM on-board to buffer incoming or outgoing packets. This on-board RAM also needs a reserved area of PC memory. This reserved area of PC memory is called the base memory address. Each card that requires a base memory address must use a unique address for proper operation. If two cards are looking for their ROM at the same memory location, neither will work. Some programs may overwrite the card's base memory and cause the workstation to hang indefinitely. You can use the Exclude parameter with EMM386 to protect a card's base memory.

Some cards allow moving the starting addresses of ROM to avoid conflicts, but a card's ROM addresses should not be changed because many software modules rely on a card's standard addresses.

The following base memory addresses are commonly used.

Base Memory Address	Assignment
00000 - 9FFFF	System memory
A0000 - B7FFF	Monochrome
B8000 - BFFFF	Super VGA
C0000 - C7FFF	CGA or Super VGA (available for ROM if not used)
D0000 - D7FFF	System memory (available for ROM if not used)
E0000 - EFFFF	AT computers
F0000 - FFFFF	System BIOS

Configuring Elements

To configure computer devices (to set up interrupt, memory address, I/O port, etc.) we use the following elements:

- DIP switches
- Jumpers
- Terminating resistors
- CMOS setup
- Device drivers

DIP Switches

DIP (Dual In-line Package) switches are used to configure expansion boards or the system board (motherboard). For example, on an Ethernet board, DIP switches are used either to configure the appropriate I/O port, or to set the board to use boot PROM (Programmable Read Only Memory). On an ARCnet board, DIP switches are used to set a node address. On the system board, they are used to configure internal hardware components like video type, amount of memory, and number of disk drives. There are two types of DIP switches: rocker and slide. A rocker switch is depressed to the ON or OFF position; a slide switch is moved to the ON or OFF side.

Jumpers

Jumpers are also used to configure computer devices. A jumper consists of two pins that stick up out of the computer board. A tiny jumper block is used to establish an electronic contact between the two pins. Jumpers are very commonly used on network boards to set IRQs, memory addresses, and I/O ports.

Terminating Resistors

Terminating resistors are used to terminate the ends of certain cable systems, including disk drives and Ethernet bus networks.

For example, if terminators are not used on both ends of an Ethernet bus network, the signal will bounce back from the end and collide with other signals.

CMOS Setup

Most computer systems store system configuration information in a separate configuration memory location. This is normally a small block of CMOS (Complementary Metal Oxide Semiconductor) ROM. The system configuration parameters include the number and type of disk drives installed, the amount of system memory, and the date and time. To keep configuration information from being lost when power is removed (that is, when the system is shut off), a battery on the system board provides power for the CMOS configuration.

During the boot sequence, the computer performs a hardware check and compares what it finds against the information stored in the CMOS. If the information matches, then normal booting proceeds. If the information is not the same, the system will display a setup configuration error.

A special BIOS setup routine can be used to configure the CMOS. This routine is usually activated by pressing the <DELETE> key during bootup.

Automatic Setup Routine

To simplify installation and avoid an accidental hardware conflict, MCA and EISA boards are designed to be configured automatically through a software setup routine and setup diskette. The software configuration program selects non-conflicting values for the boards and devices installed in the system, and writes the configuration information to the diskette and to memory. This simplifies the process of installing new devices in the system.

Most new network boards do not use DIP switches or jumpers for configuration. Instead, they come with a setup program which is used to configure the board for desired settings and to save the configuration to the board PROM.

Device Drivers

The device driver is the software that tells the operating system how to interface properly with the device. A new device driver is needed when a new piece of hardware or device is added, which is not set in BIOS. DOS has built-in drivers for the required computer components such as keyboard, monitor, and disk controllers. Any additional drivers are loaded by the CONFIG.SYS file so that the computer can recognize the new devices.

Chapter 4 Review

Q1.   What bus types are used with IBM compatible PCs?
_____
_____
_____

Q2.   What must be configured when connecting two devices?
_____
_____
_____

Q3.   What are the differences between the ISA, MCA, and EISA bus channels?
_____
_____
_____

Q4.   Define the following:

   a)   Interrupt
_____
_____
_____

b) I/O Address

c) Base Memory Address

Q5. What are the different methods of configuring a system?

Chapter 5   Computer Memory

Memory is a vital component of a computer. All computers must have a minimum amount of memory installed if they are to function properly. Data and programs (instructions for the computer) are usually stored on disks. They must be loaded into memory before the CPU can perform any operations. Computers use five different types of memory:

- RAM (Random Access Memory)
- ROM (Read Only Memory)
- PROM (Programmable Read Only Memory)
- EPROM (Erasable Programmable Read Only Memory)
- EEPROM (Electronically Erasable Programmable Read Only Memory)

Types of Memory

Random Access Memory (RAM)

RAM is the computer's main memory. Data and programs are read into RAM before any operations can be performed. The name RAM is given because any location in this memory can be accessed randomly—the memory does not have to be accessed in sequence (as it does in, for example, tapes).

RAM is only used for temporary storage of information that the CPU needs to process, so information in RAM is lost when the computer is reset or is turned off. Two different types of RAM are used in microcomputers: Dynamic RAM (DRAM), which uses capacitors, and Static RAM (SRAM), which uses switches.

Storing Data in RAM

Each character is represented by a binary number, one byte in length, whose value is determined by the particular code in use (ASCII or EBCDIC, for example). The capacitors or switches are then set to 0 or 1 in the same order that the 0s and 1s appear in the binary number. Thus, each capacitor (in DRAM) or switch (in SRAM) in a chip represents one bit of a byte. Another bit, called the parity bit, is used for error checking.

Dynamic RAM (DRAM)

DRAM contains hundreds of tiny capacitors which store information in bit format (in 0s and 1s). A charged capacitor represents a 1, and a discharged capacitor represents a 0. The problem with DRAM is that the charge in a charged capacitor is so small that it dissipates very quickly. To make up for this, DRAM is constantly recharged (refreshed) to maintain the stored information. DRAM chips are the most common memory chips used in computers. They typically feature speeds of 60 to 70 nanoseconds. A variation of DRAM is called Page Mode RAM or Fast Page Mode RAM (FPM RAM).

Extended Data Output RAM (EDO RAM)

EDO RAM is a variation of regular FPM RAM with minor changes in the data output timing. Data comes out of EDO RAM more frequently than from standard DRAM. EDO memory allows a CPU to access memory ten to fifteen percent faster than comparable FPM chips. The data comes out in 70 ns (nanoseconds), 60 ns, or 50 ns. 60 ns is the minimum requirement for a bus speed of 66 MHz. The problem with EDO RAM is that it may soon be outdated, because it hardly works with any bus speed higher than 66 MHz, which has already been reached.

Static RAM (SRAM)

SRAM uses tiny switches to store information. A switch is set ON to represent a 1, or OFF to represent a 0. The switches do not need to be refreshed to maintain the information. SRAM is faster than DRAM, but is more expensive. The speeds of SRAM range from 15 to 25 nanoseconds. This memory is referred to as the L2 Cache.

Synchronous Dynamic RAM (SDRAM)

Right now, this type of RAM seems to be the one that will be used most in the future. It is able to handle all input and output signals synchronized to the system clock, thus giving the best performance. Another good thing about SDRAM is that it can handle bus speeds of up to 100 MHz.

Parity

Regardless of which coding method is used to represent characters in RAM (main memory), it is important that the characters be stored accurately. For each byte of memory, most computers have at least one extra bit (called a parity bit) that the computer uses for error checking. A parity bit can detect a change to one of the bits in a byte. Such an error could occur because of voltage fluctuations, static electricity, or a memory chip failure.

Parity checking can be either odd or even. In computers using odd parity, the total number of ON bits (bits representing 1) in the byte, including the parity bit, must be an odd number. In computers using even parity, the total number of ON bits, including the parity bit, must be an even number. So in a computer using odd parity, a byte with four 1s in the data bits would have a parity bit of 1 for a total of five 1s (because five is an odd number). In a computer using even parity, a byte with four 1s in the data bits would have a parity bit of 0 for a total of four 1s (because four is an even number). IBM compatible PCs use even parity.

Parity is checked each time a memory location is used. When data is moved from one location to another in main memory, the parity bits of both the sending and receiving locations are compared to see if they are the same. If the system detects a difference or if the wrong number of bits is set to ON (e.g. an even number of bits in a system using odd parity), an error message is displayed. Some computers use multiple parity bits that enable them to detect multiple bit errors.

Fake Parity

With normal parity, when eight bits of data are written to DRAM, a corresponding parity bit is written at the same time. The value of the parity bit (either a 1 or a 0) is determined at the time the byte is written to DRAM, based on the number of odd and even bits in the byte as discussed above. Some manufacturers use a less expensive "fake parity" chip. This chip simply generates a 1 or a 0—whichever will satisfy the parity requirement—at the time the data is being sent to the CPU, in order to accommodate the memory controller.

For example, if the computer is using odd parity, the fake parity chip will generate a 1 whenever a byte containing an even number of 1s is sent to the CPU, and a 0 whenever a byte containing an odd number of 1s is sent to the CPU.

`This means that the fake parity chip always sends a signal that indicates uncorrupted data. This fools the computer, which is expecting a parity bit, into thinking that parity checking is taking place when no error checking actually occurs. Fake parity cannot detect an invalid data bit; it can only trick the computer into thinking that all the bits are valid.

Error Correction Code (ECC)

ECC is primarily used in high-end PCs and file servers. The important difference between ECC and parity is that ECC is capable of detecting and correcting 1-bit errors. With ECC, 1-bit error correction usually takes place without the user even being aware that an error has occurred. Depending on the type of memory controller used in the computer, ECC can also detect 2-bit, 3-bit, or 4-bit memory errors, which are much rarer. However, while it can detect these multiple-bit errors, it can only correct single-bit errors. If ECC detects a multiple-bit error, it reports a parity error.

Read-Only Memory (ROM)

ROM contains data that is permanently recorded in the memory when it is manufactured. ROM retains its contents even when the computer's power is off. The data or programs that are stored in ROM can be read and used, but cannot be altered; hence the name "read only". ROM is used to store items such as the instruction set of the computer. Instructions that are stored in ROM memory are called firmware or micro code.

BIOS (Basic Input/Output System) ROM is the most common ROM used in microcomputers. BIOS is software, stored in ROM, that works with the operating system to control the primary devices such as keyboard, disks, printers, and monitors.

Programmable Read-Only Memory (PROM)

PROM means Programmable Read Only Memory. PROM acts the same as ROM when it is part of the computer; that is, it can only be read and its contents cannot be altered. With PROM, however, the data or programs are not stored in the memory when they are manufactured. Instead, PROM can be loaded with specially selected data or programs prior to installing it in a computer.

Erasable Programmable Read-Only Memory (EPROM)

A variation of PROM is EPROM, which means Erasable Programmable Read Only Memory. In addition to being used in the same way as PROM, EPROM allows the user to erase the data stored in the memory and to store new data or programs in the memory. The data is erased by removing a protective cover and exposing the chip to ultraviolet light.

Electrically Erasable Programmable Read-Only Memory (EEPROM)

EEPROM is like EPROM in that it can be programmed and reprogrammed over and over again. The difference is that instead of using ultraviolet light to erase data as in EPROM, with an EEPROM chip an electrical pulse is used.

System Memory

So far we have discussed different types of memory and how they store information. Now we will discuss how programs or users can use this memory, and the different ways the memory can be used.

The first IBM PC used an 8086 processor, which was able to address 1 MB of memory. Figure 5-1 shows how this 1 MB was divided into three parts. The largest part is the main memory (also known as conventional memory), which is 640 KB. Program instructions and data are stored in this part of the memory. The second largest part is reserved for system ROM and is 256 KB. The smallest part, used for video memory, is 128 KB. When spoken of collectively, system ROM and video memory are referred to as Upper Memory Blocks (UMBs).

The later 80286 processor uses 16-bit addressing (instead of 8-bit) and supports 16 MB of RAM. The 32-bit 80386 and later chips (486 and Pentium) can address up to 4 GB (gigabytes) of RAM.

Three memory categories are used in a microcomputer: conventional, extended, and expanded.

Conventional Memory

DOS was designed to directly use 640 KB of RAM. The first 640 KB of memory is called conventional memory. DOS loads in this memory, as do some "helper" programs called TSRs (Terminate and Stay Resident) and device drivers. All DOS programs basically run in this 640 KB of memory. Other operating systems like OS/2, Windows NT, and NetWare do not have this limitation. Microsoft Windows and some other programs can also use memory above 640 KB.

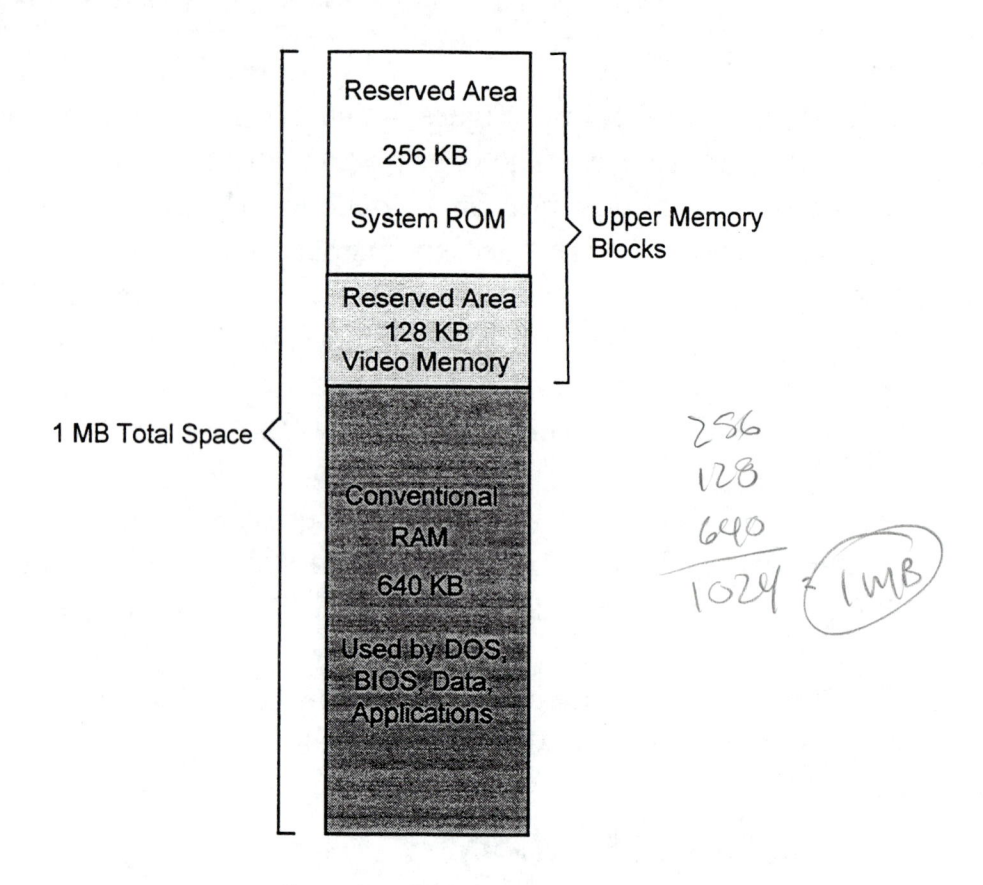

Figure 5-1 Conventional Memory

Extended Memory

Any memory in the computer beyond the 1 MB that the CPU can address is called extended memory. On 286 computers, this is the memory between 1 and 16 MB (a 286 processor can address up to 16 MB of memory), and on a 386 computer it can be up to 4 GB. 8086 and 8088 computers do not have extended memory because these processors can only address up to 1 MB of memory.

As we have seen, 286 and higher processors can support a lot more than 1 MB of memory, but DOS is still limited to 640 KB of RAM (an initial design problem). Programs like MS Windows can use extended memory.

Chapter 5: Computer Memory

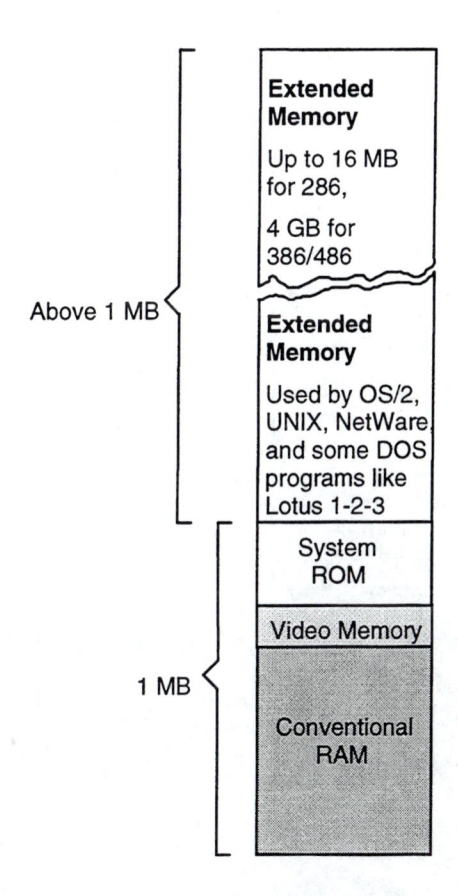

```
                    ┌─ ┌─────────────┐
                    │  │ Extended    │
                    │  │ Memory      │
                    │  │             │
                    │  │ Up to 16 MB │
                    │  │ for 286,    │
                    │  │             │
                    │  │ 4 GB for    │
                    │  │ 386/486     │
        Above 1 MB ─┤  ├─────────────┤
                    │  │ Extended    │
                    │  │ Memory      │
                    │  │             │
                    │  │ Used by OS/2│
                    │  │ UNIX,NetWare│
                    │  │ and some DOS│
                    │  │ programs    │
                    │  │ like        │
                    │  │ Lotus 1-2-3 │
                    └─ ├─────────────┤
                       │  System     │
                       │  ROM        │
                       ├─────────────┤
                       │ Video Memory│
             1 MB ────┤├─────────────┤
                       │             │
                       │ Conventional│
                       │ RAM         │
                       │             │
                       └─────────────┘
```

Figure 5-2 Extended Memory

Expanded Memory

Expanded memory (also known as paged memory) is any computer memory which is not addressable by the CPU and cannot be used directly by DOS. This memory can be used with the help of a special program called EMS (Expanded Memory Specification). The latest EMS software, LIM EMS 4.0 (developed by Lotus, Intel, and Microsoft), uses a technique called page switching to swap different "pages" of information, usually 16 KB in size, in and out of the area of 256 KB reserved for system ROM. As we discussed before, this reserved area is part of the 1 MB RAM which is directly addressable by the CPU.

Expanded memory is limited to 32 MB in any system. Expanded memory can be used in 8086/8088, 286, 386, and above computers.

```
┌─────────────────┐      ┌─────────────┐
│ Reserved Area   │      │  Expanded   │
│     256 KB      │      │   Memory    │
│  System ROM     │      │             │
│  and buffers,   │      │ Up to 32 MB │
│  including LIM  │      │             │
│  page buffers   │      │             │
├─────────────────┤      └─────────────┘
│ Reserved Area   │
│     128 KB      │      ┌─────────────┐
│  Video Memory   │      │  Expanded   │
├─────────────────┤      │   Memory    │
│                 │      │             │
│  Conventional   │      │ Used by some│
│      RAM        │      │ DOS programs│
│     640 KB      │      │             │
│                 │      │ Arranged into│
│  Used by DOS,   │      │ 16-KB "pages"│
│   BIOS, Data,   │      │ of memory and│
│  Applications   │      │ does not have│
│                 │      │  addresses  │
└─────────────────┘      └─────────────┘
```

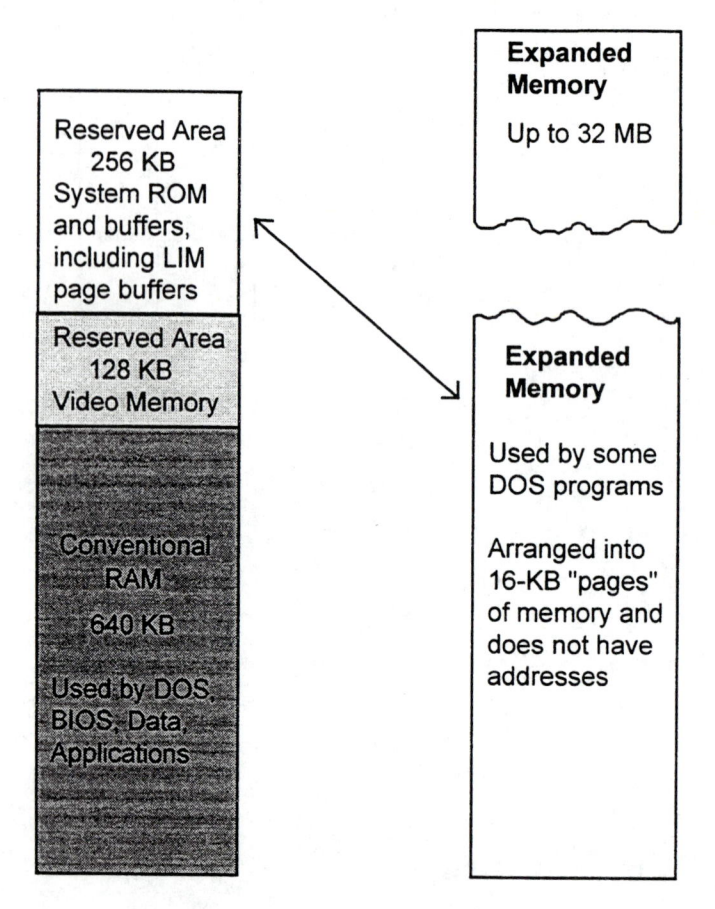

Figure 5-3 Expanded Memory

Some portions of PC memory are also referred to as upper memory and high memory.

Upper Memory

This is the 384 KB of memory between 640 KB and the real mode boundary of 1 MB (1024KB). This area is used for special purposes such as ROM BIOS, serial ports, adapter cards, and video. In Microsoft DOS the HIMEM.SYS device driver works in conjunction with EMM386.SYS, the expanded memory manager, to provide UMBs (Upper Memory Blocks) which can be used for device drivers and memory resident programs.

High Memory Area (HMA)

High memory (also known as High Memory Area) comes from a bug in the design of the 286 and later chips. It turns out that these chips can not only access the 1024 KB of memory in Real mode, but can actually access another 64 KB, which raises the limit a little bit to 1088 KB. That extra 64 KB is taken from extended memory.

Not everything can be loaded into the HMA, even if it is small enough. This is why DOS 5.0 was such an improvement over earlier versions of DOS: 5.0 and later versions can actually fit most of themselves inside the HMA, leaving more conventional memory for normal DOS programs.

Direct Memory Access (DMA)

DMA, discussed in Chapter 4, uses an additional chip to create a path between peripherals and main memory which bypasses the CPU. The CPU is thereby freed from simple but time-consuming data transfers, and can execute other processes.

Chapter 5 Review

Q1. What are the main types of RAM?
_____
_____
_____

Q2. What is a common use of ROM?
_____
_____
_____

Q3. What is a parity bit?
_____
_____
_____

Chapter 6   Auxiliary Memory— Storage Devices

Auxiliary memory refers to physical means of storing programs and data other than the primary computer memory (RAM and ROM). Disks are the most common computer data storage medium. Other means are also frequently used, including CD-ROMs, tapes, removable media, and DVD disks.

Disks

Computer disks use the same means of recording data as audio or video tapes. The disk is coated with a special recording medium. Small particles on this medium are magnetized by a recording or read/write head to represent 1s. Demagnetized spots represent 0s. There are two types of disks: floppy disks and hard disks.

Floppy Disks

Floppy disks, also called diskettes, are widely used for computer data storage. The IBM XT and most early PCs accommodate 5.25" disks. Newer PCs are typically equipped to handle 3.5" disks.

Floppy disks (and hard disks) must be formatted before they can store data. When you format a disk, the operating system prepares the disk so it can store the data. Formatting checks the disk surface for damage and imperfections, divides and organizes the disk into electromagnetic divisions called tracks and sectors, and creates a File Allocation Table (FAT) to store the location of files on the disk. Files are stored track by track and sector by sector. A set of contiguous sectors is called a cluster. Files are not always stored in contiguous sectors. When files on a disk are not stored in contiguous sectors, that disk is considered fragmented.

Disk Capacity and Density

The following table shows the relationship between floppy disk size, capacity, and density.

Size	Capacity	Density/Sides
5.25"	360 KB	double-density, double-sided
5.25"	1.2 MB	high-density, double-sided
3.5"	720 KB	double-density, double-sided
3.5"	1.44 MB	high-density, double-sided

The capacity of a disk is measured in either kilobytes (thousands of bytes) or megabytes (millions of bytes). A 5.25-inch high capacity disk drive can read and write both 1.2 MB disks and 360 KB disks, and a 3.5-inch high capacity disk drive can read and write both 1.44 MB disks and 720 KB disks. But a 5.25-inch low capacity disk drive can only read and write 360 KB disks, and a 3.5-inch low capacity disk drive can only read and write 720 KB disks. A drive can read disks of lower capacity, but not higher.

The density of a disk refers to the amount of information that can be stored on the disk. Data stored on disks can be lost if the disk is physically damaged or exposed to magnetic fields.

Hard Disks

Hard disks provide larger and faster auxiliary storage capabilities for personal computers. Hard disks consist of one or more rigid metal platters coated with a metal oxide material that allows data to be magnetically recorded on the surface of the platters. Each platter has two read/write heads, one above and one below. The hard disk is much faster than a floppy disk.

Like floppy disks, data on a hard disk is organized into tracks and sectors. Hard disks also use cylinders to organize data. A cylinder is made up of all the tracks on all the disk platters that lie in the same relative position. Cylinders are numbered according to the track they reference. For example, the outermost track is called track 0, and all the track 0s on all the platters are collectively referred to as cylinder 0. Hard disks are available in capacity sizes ranging from 10 MB to over 9 GB (9,000 MB).

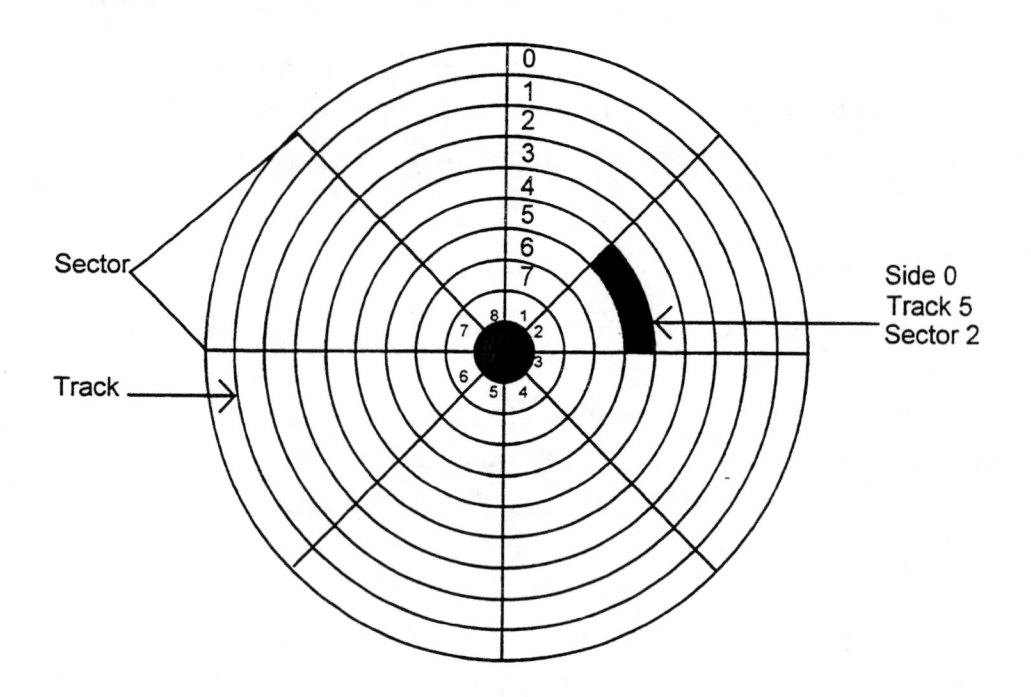

Figure 6-1 Formatted Disk Surface Showing Tracks and Sectors

Partitioning the Hard Disk

Creating logical areas in a hard disk is called partitioning. A hard disk must be partitioned before it can be formatted. Multiple logical partitions can be created on a single physical disk. To DOS, each partition appears to be a separate, logical, unique hard disk. During partitioning, an active logical partition must be identified. This is the partition that will be used to boot the system, so it is referred to as the bootable partition or the primary partition. After the primary partition has been defined, all other partitions are referred to as extended partitions.

In MS DOS 3.3, and all previous versions of DOS, you are limited to partitions of up to 32MB. Later DOS versions allow much larger partitions. If the DOS version restricts your partition to a size less than the capacity of your disk, you can create multiple partitions. Also, when you want to use two or more operating systems, you can create a separate partition where each operating system will store its own file system. The DOS FDISK program invokes a menu that is used to create logical partitions.

An extended DOS partition cannot be deleted while a logical drive is defined in that partition.

Random Access Time

The Random Access Time is the average amount of time required for the drive to deliver data after the computer sends a data request. Hard drive manufacturers often advertise their products by access time, usually in milliseconds. The transfer rate for the computer system may be slower, and therefore would be the factor that limits the computer speed.

A head is an electromagnetic device used to read and write to and from magnetic media such as hard and floppy disks, tape drives, and compact disks. The head converts the information into electrical pulses that are sent to the computer for processing. The term HDI (Head to Disk in Interference) relates to how well the head does its job.

Hard Disk Controllers

The hard disk drive is controlled by the disk controller board. Most fixed disk controllers can support up to 2 fixed disk drives. Many newer systems have the drive controller embedded directly in the motherboard circuitry, so a separate controller is not required. It is important that the hard disk and the controller use the same interface; otherwise, they will not be able to work together. This interface is used by the controller to transfer data between the disk and the CPU. There are four disk interface standards:

- ST506/412
- ESDI
- SCSI
- IDE

ST506/412

ST506/412 was developed by Seagate Technology in the late 1970s. It has the following features:

- Data transfer rate of 5 megabits per second (Mbps)
- Choice of data encoding methods: Modified Frequency Modulation (MFM) or Run Length Limited (RLL)

The RLL data encoding method was developed later. It increases the data transfer rate to 7.5 Mbps and doubles the disk capacity. The ST506/412 standard can use either RLL or MFM, but a disk controller created for MFM cannot be used with RLL.

Enhanced Small Device Interface (ESDI)

ESDI is an enhanced version of ST506/412. It was developed in the mid-1980s. ESDI is accepted as a standard by the American National Standards Institute (ANSI). The two main enhancements that ESDI provides are:

- Data transfer rate of 10 Mbps, or even 15 Mbps with some faster controllers
- Higher storage capacity

Both ST506/412 and ESDI are serial interfaces, meaning that they transfer data from the drive to the controller one bit at a time.

Small Computer System Interface (SCSI)

SCSI (pronounced "scuzzy") is a very fast interface which uses parallel communication (that is, it transfers 8 bits at a time). Instead of providing a drive interface like ESDI, SCSI provides a system-level interface. A system-level interface allows the computer to connect to (or support) additional storage devices, including CD-ROMs, tape drives, read/write optical drives, and so on. Up to 7 devices can be attached to one SCSI controller. This standard is currently the first choice for larger server disks.

Integrated Drive Electronics (IDE) Interface

IDE was developed by Western Digital Corporation. IDE drives have controller electronics on the drive itself, which makes them reliable and also frees up an expansion slot which other interfaces use for the controller. IDE drives use RLL encoding and operate at data rates of 7.5 Mbps to 10 Mbps. Most computers used as workstations have IDE drives.

Other Storage Devices

CD-ROM

CD-ROM disks are flat disks, identical in size to those played in CD audio players. They typically feature slow access, about 100 milliseconds, although some newer CD-ROMs are faster than this. The transfer rate of a CD-ROM drive is the factor that limits its speed. This rate is expressed in "time" (X), where 1X is equivalent to 150 KB per second.

Therefore:

10X = 1.5 MB/sec (1500 KB/sec)
20X = 3.0 MB/sec
24X = 3.5 MB/sec

The maximum capacity of a CD-ROM disk is 682 MB.

Tape Backup

Tape backup systems—most often QIC (quarter inch cartridge)—are frequently used as an inexpensive means to regularly backup programs and data. The backup systems usually compress the data. The disadvantages to tape backup are the necessity to go through the entire tape to access a small amount of data, and the fact that the tapes are not considered very reliable.

Backup Strategies

Backups are vital in preventing data loss. In a backup, a copy of the data is made and is then stored in a different location. Without this management step, a disk failure could mean the permanent loss of data. When regular backups are made, data loss is minimized because the current version of any file not created or modified since the last backup can be recovered, and even modified files can have their earlier versions restored.

The most reliable backup strategy would be to back up every file when it is created and every time it is modified. However, this would be costly in both time and materials. Instead, three less comprehensive backup methods are used:

- Full
- Incremental
- Differential

Full Backup

In full backup, all data is backed up on a periodic basis. This is the basic backup method, and should always be part of any backup strategy. One of the other two methods is then selected to be used between full backups. (Only one of the other two methods is chosen because they use different means of selecting files to back up.) Depending on the amount of system activity, the time between full backups can vary from less than a week for a busy system to several months for an inactive system.

Many operating systems supply a file attribute, called the archive bit, which is used by backup programs to determine whether a file needs to be backed up. The archive bit indicates when a file is new or modified (and therefore needs to be backed up) by setting its value to 1. The backup process may then clear the archive bit (change its value to 0) or leave it set. Full backups normally clear the archive bit.

Incremental Backup

This method will periodically back up only the files shown by the archive bit to have been created or modified since the last backup (either full or incremental). After saving the file, the bit is cleared. This method is faster than either full or differential backup because it backs up only new and modified files, and does nothing with unchanged files. All incremental backups between two full backups should be kept, as they will all be needed if a data loss occurs. This means that incremental backup requires as many separate storage units (disks or tapes) as there will be backups.

Differential Backup

This method also selects only the files shown by the archive bit to be new or modified, but in differential backup the archive bit is not cleared after the file is saved. This means that the backup program will select all files created or modified since the last full backup, not just since the last differential backup. For this reason, it is slower than incremental backup, though still faster than full backup, and it needs less storage units than incremental backup because only the last backup needs to be kept.

Restore Strategies

If it is necessary to restore the system data from the backup, the last full backup will be used first. Then one or more additional backups will be required, depending on whether you have employed the incremental or differential strategy.

If incremental backups were made, then all the backups must be used in the order in which they were made, making the restoration process long and complex. If differential backups were made, then only the last backup is needed, so restoration is faster. This is the reverse of the situation in backing up, where incremental is faster than differential.

Chapter 6 Review

Q1.   What are the two main types of floppy disks?
_____
_____
_____

Q2.   How many bytes can be stored on a 3.5-inch high-density disk?
_____
_____
_____

Q3.   For each of the following, state whether it is a disk interface or a data encoding method.

a)   IDE   _____
b)   MFM   _____
c)   ESDI   _____
d)   RLL   _____
e)   SCSI   _____
f)   ST506/412 _____

Chapter 7 Video Display Unit

The microcomputer's main form of output is the visual display unit (VDU). Most VDUs have a cathode ray tube (CRT) similar to that in a television set. In fact, with many personal and home computers we can use a television as the VDU. Portable and laptop computers use a flat screen with liquid crystal or gas plasma techniques instead of a CRT.

VDU Features

There are two basic modes of video display operation: text and graphics. The text mode is used to display basic characters. The graphics mode displays a variety of colors and images. Three basic display elements define the graphics mode: quality, resolution, and color.

Display Quality

The top quality of a display unit has clearly defined characters and sharp graphics.

Pixels and Resolution

The term "pixel" is short for "picture element". A pixel is the smallest addressable element in an electronic display: a single dot on the screen. These dots are combined to create images. The number of pixels per square inch of screen space is called resolution. The better the resolution, the clearer the picture.

Color

Monochrome screens specially designed for use with computers usually display a single text color, such as white, green, or amber, on a black background. The characters are displayed without flicker and with very good resolution. Some monochrome screens have graphics capabilities.

The use of color screens is increasing in business and science applications, because numerous studies have found that color enables the user to more easily read and understand the information displayed on the screen.

Monitor Types and Displays

There is an important difference between a display and a monitor. The two terms are often used interchangeably, but the display is the device or the technology that produces the image on the screen. The monitor is the device which includes both the display itself and the display electronics. The CRT technology used in television sets and monitors uses either a digital or an analog signal in producing the video image. Analog monitors produce an image that is much higher in resolution, with a greater variety of colors, than digital monitors.

Monochrome Monitors

Monochrome monitors can display a single color on a black background. There are four basic monochrome monitors, each with its own adapter:

- TTL
- Composite Monochrome
- VGA Monochrome
- Multiscanning Monochrome

TTL (Transistor-Transistor Logic)

TTL, the original monitor for the first IBM PC computer, uses digital signals generated by the TTL family of integrated circuits.

Composite Monochrome

This monitor has the lowest resolution of all the monochrome monitors. It has the same resolution as the CGA, discussed below, and can use a CGA or CGA compatible adapter.

VGA Monochrome

This monitor uses an analog video graphics array signal, but is incompatible with color VGA video standards.

Multiscanning Monochrome

The multiscanning monitor can use signals from all other monochrome adapters. Multiscanning means that the monitor automatically switches over to the frequency and standard required by the adapter.

Color Monitors

There are five major color display types:

- RGB
- CGA
- EGA
- VGA
- Multiscanning color displays

RGB

The name RGB comes from the three primary colors: red, green, and blue. These colors are added to each other on the screen to produce others. RGB was the first color display for the IBM PC.

CGA (Color Graphics Adapter)

CGA was the first color graphics standard. It displays 16 colors and features text, graphics, and color modes. It can work with monochrome displays. CGA provides a very low resolution, approximately 320x200 dots per inch (dpi).

EGA (Enhanced Graphics Adapter)

EGA is an enhanced form of RGB. It has a resolution of 640x350 dots per inch and can display up to 64 colors. EGA can work with a monochrome screen to allow graphics capabilities.

VGA (Video Graphics Array)

VGA is an enhancement of EGA. It has a resolution of 320x200 dots per inch, and in this resolution it can display up to 256 colors. In higher resolutions, such as 640x480, it can display only 16 colors. Super VGA can display up to 256 colors with a resolution of 800x600.

Multiscanning Color Displays

One of the first and most successful multiscanning systems was the NEC Multisync. Multiscanning displays can work with almost any video standard.

Chapter 7 Review

Q1. Name two monochrome monitors.
_____
_____
_____

Q2. Which color monitor has the greatest resolution?
_____
_____
_____

Chapter 8   Input/Output Ports

The input/output ports are used to provide an interface between the computer and other external devices like printers and modems. Here we will discuss two types of ports: parallel and serial.

Parallel Ports

Parallel ports transfer data eight bits (one byte) at a time by using eight different wires: each bit has its own wire, and all the bits in one byte move at the same time. Parallel ports are used almost exclusively with printers. A parallel printer cable has a 25-pin D connector on one end (which connects to the computer) and a 36-pin card edge connector on the other (which connects to the printer). Parallel ports are typically defined as LPT1, LPT2, and LPT3. LPT stands for "local printer", and there are a maximum of three parallel ports in an IBM compatible PC. The main advantage of parallel communication is that data can move eight times faster than if it had to flow one bit at a time down a single wire.

The biggest problem with parallel communication is crosstalk. Crosstalk is a condition which occurs when parallel signals travel long distances and two different signals interfere with each other. The longer the cable, the greater the effect of crosstalk. Crosstalk can be reduced to a point where it has no effect on data integrity simply by shortening the cable length. Most manufacturers recommend that parallel connections be kept under 10 feet to prevent problems. Computers and printers vary greatly in their sensitivity to crosstalk. Some systems may work with parallel connections up to 50 feet long, while others must be kept under 10 feet.

Another disadvantage of parallel communication is that input/output chips in older systems are unable to receive data, so they can only control unidirectional parallel devices—that is, these chips allow only one-way communication. This means that the external device can only receive information from the processor; it cannot transmit any information to the processor. Newer input/output chips can receive data and are bi-directional.

Serial Ports

The serial port transfers one bit at a time over a single wire. This communication is not as fast as parallel communication, but it is more reliable and can be used over much longer distances.

The serial port is also called an asynchronous port or COM port. This is short for communications interface port, which it is called because it is used to connect a modem to a computer. The port operates according to the EIA (Electronics Industry Associations) RS232C standard, so it is also known as a RS232 port. IBM PCs can have up to four serial ports (designated as COM1, COM2, COM3, and COM4), but you can only use two serial devices at the same time.

Serial Device Types

The RS-232 standard defines two different types of devices: Data Terminal Equipment (DTE) and Data Communication Equipment (DCE). A DTE is a device that a user interfaces with, usually a terminal or computer. A DCE connects a DTE to the communication medium. A modem is a example of a DCE which converts the DTE's data format to a signal suitable for the medium.

Communication Synchronization Methods

Two methods are used to keep track of the bits as they travel from one device to another: asynchronous communication and synchronous communication.

Asynchronous Communication

In this method, start and stop bits are used with each character (a series of bits, usually between five and eight) for bit synchronization. The transmitter sends a start bit to begin transmission. When the receiver identifies a start bit, it begins an internal clock and continues measuring the signal at predetermined bit intervals. This process continues until the transmitter sends a stop bit. A parity bit is sometimes included at the end of each character to provide error checking.

In the asynchronous synchronization method, both transmitter and receiver use similar timing but their clocks are not synchronized with each other.

This method is used for lower-speed data transmission, and is used with most communication equipment designed for personal computers.

Synchronous Communication

In this method, blocks of bits are transmitted instead of one character at a time. The transmitter and receiver maintain synchronization with the help of an external clocking mechanism that sends timing signals on a separate line or wire. This synchronizes the devices' clocks to the correct bit timing. Included in the data are error checking bits and sync bytes (start and end indicators).

Synchronous transmission requires more sophisticated and expensive equipment than asynchronous transmission, but gives much higher speed and accuracy.

Bps and Baud Rate

The transmission rate of a communication channel is determined by its bandwidth and its speed. The bandwidth is the range of frequencies that a channel can carry. Since transmitted data can be assigned to different frequencies, a wider bandwidth that offers more frequencies means that more data can be transmitted at the same time. The speed at which data is transmitted is usually expressed as bits per second or as a baud rate.

Bits per second (Bps) refers to the number of bits that can be transmitted in one second. Using a 10-bit byte to represent a character (7 data bits, 1 start bit, 1 stop bit, and 1 parity bit), a 2400 bps transmission would transmit 240 characters per second. At this rate, a 20-page single-spaced report would be transmitted in approximately five minutes.

Baud rate refers to the number of times per second that the transmitted signal changes; for example, the speed of the oscillation of the sound wave on which a bit of data is carried over a phone line.

At speeds up to 2400 bps, usually only one bit is transmitted per signal change and thus the bits per second and the baud rate are the same. To achieve speeds in excess of 2400 bps, compression and encoding are used to allow more than one bit of data to be sent with each signal change. Therefore the throughput can be greater than the baud rate. The data rate (throughput) is measured in bps, and the communication rate is measured in baud.

Handshaking

Handshaking is a process through which DTEs and DCEs prepare to transmit data. For example, when two DTEs (for example, two computers) want to exchange data via two DCEs (for example, two modems), they start the handshaking process by applying power and sending signals to different pins of RS-232 connectors. Handshaking can be accomplished by either hardware or software.

Flow Control

Serial communication is two-way communication, so some more active methods can be implemented to ensure error-free transmission. Flow control is one of these methods. Flow control allows the receiving device to send a signal to the sending device, indicating whether it is ready to receive data or it needs the sender to wait. Flow control can be accomplished by either hardware or software.

One of the most common methods of software flow control is XON/XOFF. The XON/XOFF software in the receiving device monitors how much data is coming in and how fast it is coming in. When the data buffer is nearly full, the receiving device sends an ASCII character (19, <CTRL S>) which tells the sending device to temporarily stop transmitting. When the data buffer has emptied out again, the receiving device sends an ASCII character (17, <CTRL Q>) to tell the sending device to resume transmitting.

Chapter 8 Review

Q1. How many parallel ports are allowed by IBM?

Q2. What is a common flow control method?

Q2. What are the different serial transmission methods?

Unit 2: DOS for Network Users

Chapter 9 Disk Operating System (DOS)

The term DOS stands for Disk Operating System. DOS provides an interface between software programs and the microcomputer components. The main function of DOS is to manage the flow of information through the computer system.

Components of DOS

DOS has two main components:

- DOS Core Operating System
- Auxiliary Utility Programs

Each part is made up of several files.

DOS Core Operating System

The core operating system consists of three programs, which are loaded into the computer's main memory and reside there until the computer is turned off or is reset. These three programs are:

- IBMBIO.COM
- IBMDOS.COM
- COMMAND.COM

IBMBIO.COM

IBMBIO.COM (also known as IO.SYS) handles the basic input and output devices such as the keyboard and the video display unit (monitor).

IBMDOS.COM

IBMDOS.COM (also known as MSDOS.SYS) is the central program that controls the internal communication between the microcomputer components.

Note: IBMBIO.COM and IBMDOS.COM are hidden files.

COMMAND.COM

COMMAND.COM (also called the command interpreter) interprets or executes all the commands entered at the keyboard, and redirects the requests to CPU, disk, memory, and other components.

Auxiliary Utility Programs

These utility programs can be divided into two groups: Internal and External. Internal commands are part of the Core Operating System and are contained within the COMMAND.COM file. External commands are separate files (programs), kept under the DOS directory, with extensions of .EXE or .COM.

The Computer Startup Process (Boot)

DOS programs are normally stored on a diskette or on a hard disk. To begin using the operating system, the DOS programs must be read into main memory, a process known as booting. On a system with two disk drives, you insert the disk containing the DOS into the A: drive of the computer and turn on the computer. On a system with a hard disk, DOS is already available on the hard disk. If the hard disk is not functioning, or if you are installing a new hard disk, then you boot from the A: drive.

Cold Boot Process

Starting the computer by turning on the power switch is known as a cold boot or cold start. The computer will first run some tests to diagnose its own circuitry. After running these tests, the computer runs the BOOTSTRAP loader program from ROM to begin loading DOS. The BOOTSTRAP loader program is stored in the BIOS chip. It interacts with CMOS (Complementary Metal Oxide Semiconductor, discussed later) to gather hardware information before loading the operating system.

Warm Boot Process

Restarting the operating system by pressing the CTRL, ALT, and DEL keys simultaneously or by pressing the Restart button (if there is one) is called a warm boot or warm start. This procedure does not repeat the initial diagnostic test, but it erases all programs and data from main memory and reloads DOS.

The term "boot" comes from the name of the BOOTSTRAP loader program. In a microcomputer, this program is located in permanent memory storage called the ROM-BIOS (Read Only Memory - Basic Input/Output System). When DOS boots, the system performs the following functions:

- Powering ON your computer
- DOS search
- DOS configuration

Powering ON Your Computer

Whenever power is supplied to the computer, the central processing unit (CPU) first checks its own circuitry to make sure it is functional. It will then look for the BOOTSTRAP loader and perform any instructions that are present. The boot program causes the CPU to check all the other computer equipment.

DOS Search

The BOOTSTRAP loader then causes the CPU to look for IBMBIO.COM and IBMDOS.COM (IO.SYS and MSDOS.SYS when using DR DOS or MS DOS), first in the A: drive and then in the C: drive. If the CPU finds these two files in either of these two drives, these files are immediately loaded into memory and the DOS operating system takes control. If these files are not found, the CPU displays an error message on the monitor: "Missing Operating System".

DOS Configuration

After the two DOS files (IBMBIO.COM and IBMDOS.COM) are loaded into the memory, CPU looks for a configuration file named CONFIG.SYS. This file contains the computer operating variables. After this the CPU loads the command interpreter (COMMAND.COM). After COMMAND.COM takes control, it causes the CPU to look for the AUTOEXEC.BAT file. If an AUTOEXEC.BAT file is present, the commands in this file are executed.

The user creates the CONFIG.SYS and AUTOEXEC.BAT files, and can modify these files to configure the computer.

Chapter 10   DOS Internal Commands

Using DOS Commands

The following general rules should be followed when using DOS:

- Commands can be entered in either uppercase or lowercase letters.
- Always include a colon when specifying a drive letter.
- To abort a command, press <CTRL C> (that is, press the Control key and the C key simultaneously).

F1 and F3 Function Keys

These two function keys are used to save keystrokes. They are both used to reproduce the last DOS command entered. The F1 key results in one character at a time being added to the line with the command prompt. The F3 key enters the previous command from the current position to the end of the line.

HELP and FASTHELP

These programs provide help and documentation for the DOS commands and related topics. In the help files, there is a discussion of the command, its required syntax, an explanation of the syntax, notes and examples. The syntax is:

HELP *[command]*
   or
FASTHELP *[command]*
   or
command /?

(Unless specified otherwise, whenever an example of syntax is given in this book, *italicized* text indicates a substitution while *[bracketed]* text indicates an optional parameter.)

The HELP command without any parameters will provide a help menu, which DOS calls a command reference. Commands and related topics can be selected to obtain help.

Command Parameters

Command parameters are also known as switches. When a command is typed, additional parameters may need to be specified to further define the function that will be invoked. A parameter is usually placed on the command line after the command name. It is most often a slash (/) followed by a letter or a variable. For example, the DIR command used with a parameter would look like this:

DIR /W

Not all DOS commands require parameters.

Syntax

As with any language, DOS uses specific syntax. Also, unlike humans, computers are unable to understand text with any mistakes—no matter how small—in spelling, spacing, and punctuation (slashes, spaces, brackets, etc.). Therefore, it is very important for you to check the spelling and syntax before invoking any command with the <ENTER> key, and especially when editing configuration and batch files.

The DOS Command Processor (COMMAND.COM)

DOS commands are processed by the COMMAND.COM program, which is also known as the DOS command interpreter or command processor. When a command is entered, the COMMAND.COM program determines whether the command is already part of the COMMAND.COM program. If it is, the command is known as a DOS internal command, and the command is executed immediately. If the command is not internal to COMMAND.COM, it must be found outside of the COMMAND.COM program. Such commands are known as external commands, and are actually individual programs that are kept on disk.

Internal DOS Commands

Internal commands are part of the operating system COMMAND.COM file. Once we have loaded DOS into main memory, internal commands are available. We can enter an internal command at any time. It does not matter whether the DOS system diskette is in the default drive. DIR, COPY, CLS, ERASE, RENAME, and DEL are some examples of internal commands.

CLS

The CLS command clears the screen.

Syntax:

CLS

VOL

The VOL command allows you to display the volume label of the disk at any time. A volume label is an internal label that DOS places on the disk. VOL does not allow you to change a label. For that, you must use the LABEL command.

Syntax:

VOL *[drive:]*

VER

VER is short for "version". If you want to display the DOS version number, use the VER command.

Syntax:

VER

DIR

The DIR command lists the files in the current directory. It does not display hidden files. Wildcard characters and other parameters can be used when listing files.

The DIR command lists the following information:

- Volume Label – Appears at the top of the list. Volume labels identify the disk.
- Directory Path – Displays the drive letter and the directory path name.
- File Name and Extension – The extension identifies the type of file.
- File Size – Displays the number of bytes contained in the file.
- Date and Time – Displays the last time the file was modified.
- Number of Files and Free Space – Displays the number of files in the list and the total number of free bytes remaining on the disk.

Syntax:

DIR *[drive:][path][filename][/switches]*

The three switches that are commonly used with the DIR command are:

- /P – causes the command to pause after the screen is filled (23 entries).
- /W – causes DOS to present the files in five columns across the screen.
- /O – only used in DOS 6.0 or above, it lists the directories and files in alphabetical order.

DATE and TIME

When we enter the DATE or TIME command, the computer displays the date or time stored in the computer. If the date or time is correct we can accept it by pressing the <ENTER> key. If it is not, we can type in a new time or new date. The time can also be entered in the 24-hour clock format. Only the hours have to be entered. The minutes and seconds are optional.

When entering the date or time, you must observe the following guidelines:

- You can use either hyphens (-) or slashes (/) to separate the month, day, and year.
- Use colons (:) to separate the hours, minutes, and seconds.
- You do not have to specify the century in the year.

Syntax:

DATE *mm-dd-yy*
TIME *hours[:minutes:seconds [a/p]]*

PROMPT

A cursor is a moveable marker on the screen that indicates the position where the next character will be entered or deleted. Preceding the cursor is the DOS prompt, which may be referred to as the system prompt or the command prompt. The appearance of the DOS prompt is controlled by the PROMPT command. The syntax of the PROMPT command is:

PROMPT $*parameters*

(If you don't know what parameter(s) you should enter, type HELP PROMPT at the command line.)

The PROMPT parameters are:

- $p – the drive letter and full path of your current directory
- $t – the current time
- $d – the current date
- $v – the DOS version number
- $n – the default drive
- $g – a greater-than sign (>)
- $q – an equals sign (=)
- $$ – a dollar sign ($)
- $_ – a carriage return or line feed

Note: You can also display any text by typing the text after the prompt command. For example, if you type

PROMPT What?

then your DOS prompt will be:

What?

The most widely used PROMPT parameters are PG. If you use this syntax:

PROMPT pg

then your prompt will display the current directory and a greater-than sign, like this:

C:\>

Typing the PROMPT command with no parameters will reset the system prompt back to the drive letter and greater-than sign, like this:

C>

MD (Make Directory)

To create a subdirectory, use the MD (or MKDIR) command. For example, to create a subdirectory called PCDOS on a diskette, type MD A:\PCDOS and press the <ENTER> key.

Directory names follow the same rules as files. A name can be up to 8 characters, followed by an extension of up to 3 characters. The characters allowed are also the same.

Syntax:

MD *drive: path*

CD (Change Directory)

You can use the CD (or CHDIR) command to move from one directory to another. Enter the letters CD immediately following the DOS prompt. Then type the backslash character and the subdirectory name, and press the <ENTER> key. To return to the root directory, simply type CD\ and press the <ENTER> key. The backslash character entered by itself signifies the root directory.

CD *[drive:]path*

RD (Remove Directory)

The Remove Directory (RD or RMDIR) command deletes directories that are no longer needed. There are two things to be careful of when using this command:

- The directory must be empty (that is, it must contain no files or subdirectories) before you attempt to remove it. DOS will warn you if it is not.
- It is not possible to remove the root directory, your current directory, or any directory between your current directory and the root directory.

Syntax:

RD *drive: path*

COPY

The COPY command copies files or a group of files—either from one directory to another on the same disk, or to another disk. It can also duplicate a file on the same disk while giving it a different file name. When using the COPY command, you type the location and name of the file you want to copy from, followed by the location and name of the file you want to copy to. The first file is called the SOURCE file and the second file is called the DESTINATION file.

Syntax:

COPY *sourcefilename destinationfilename*

REN (Rename)

The REN (RENAME) command changes the name of a file without copying it or changing its location. This command is especially helpful for organizing files. For example, suppose you have two versions of a file named PRICES.LST. The version on the disk in drive A: contains last year's prices, whereas the version on drive C: is current. To avoid confusion between the two files, you can use the REN command to rename the file that contains outdated prices. REN cannot be used to rename directories; for that, you must use the MOVE command.

Syntax:

REN *[drive:][path] oldfilename newfilename*

DEL (Delete) and ERASE

The DEL and ERASE commands do exactly the same thing: delete a file or a group of files. To delete a group of files, use the DEL command with one or more wildcards. Before using wildcards to delete a group of files, it is helpful to use the DIR command to determine what files the wildcards might delete. The switch /p prompts for confirmation.

Syntax:

DEL *[drive:][path] filename [/switches]*

TYPE

The TYPE command is used to view the contents of text files and batch programs. Text files are displayed in alphanumeric characters. Machine-readable files (like COMMAND.COM) are composed of machine language symbols. We can look at the contents of other kinds of files, too. However, only the text is readable. When you use the TYPE command, DOS displays the entire file on your screen. You cannot change the text, and you cannot view only a portion of the file. Using the TYPE command on a file with a .COM or .EXE extension produces a series of unreadable symbols.

Syntax:

TYPE *[drive:][path] filename*

PATH

The PATH command allows DOS to search through several directories looking for a specific file to complete the command or to carry out the requested task. All directories listed in the path statement must physically reside on the disk. PATH statements usually specify several directory locations; when this is the case, the locations must be separated by semicolons (;). PATH can also be used without parameters, in which case it displays the current path.

Syntax:

PATH *[drive1:][path1]*; *[drive2:][path2]*; ... *[driveN:][pathN]*

Example:

PATH C:\DOS; C:\WINDOWS

Chapter 11   DOS External Commands

External commands are stored on disk as program files. They must be read from the disk into main memory before they can be executed. FORMAT and CHKDSK are examples of external commands. Another easy way to identify external commands is to look for a .COM or .EXE extension following the file name.

Example: TREE.COM.

ATTRIB

An attribute controls access to a file. The ATTRIB command displays and changes file attributes. For example, making a file read-only prevents it from being modified or erased.

The following command options control the read-only attribute:

- +R – sets a file read-only
- -R – disables the read-only mode

The archive attribute is used to mark files that have changed since they were previously backed up. The following command options set or reset the archive flag:

- +A – sets the archive flag
- -A – clears the archive flag

The shared attribute means that a file can be used by multiple users simultaneously. This attribute is normally assigned to application files. The following command options control the shared attribute:

- +S – sets a file shared
- -S – clears the shared flag

Setting the hidden attribute on a file hides that file from the user. A hidden file will not be listed if the DIR command is used.

- +H – sets the hidden flag
- -H – disables the hidden mode

The /s switch processes files in the current directory and all of its subdirectories.

Syntax:

ATTRIB *[±R][±A][±S][±H] [drive:][path:][filename(s)][/s]*

To display the attributes of all the files in the current directory, use the following syntax:

ATTRIB

LABEL

Unlike the VOL command, the LABEL command makes it possible to create, change, or delete a volume label after the disk has been formatted. The volume label can be up to 11 characters, including spaces. This label is displayed whenever you view the directory.

Syntax:

LABEL *[drive:][label]*

Volume labels identify the disk to the computer and are used by programs to verify the proper disk.

EDIT

Edit is a basic text editor program. It requires two programs: EDIT.COM and QBASIC.EXE. It is frequently used to create and edit text files, including configuration files and batch files.

Syntax:

EDIT *[drive:][path][filename]*

MOVE

This utility is used to rename directories, and to move files to a different directory. By default, an existing file will be overwritten by a new file with the same name. The /Y switch forces these overwrites to take place without prompting the user for confirmation. The /-Y switch forces DOS to prompt for confirmation before any such overwrite.

Syntax:

MOVE *[/Y\-Y][drive:][path]filename [drive:][path]filename*

DISKCOPY

This command copies the entire contents of the floppy disk in the source drive to a formatted or unformatted floppy disk in the destination drive. DISKCOPY destroys the existing contents of the destination disk as it copies the new information to it. DISKCOPY cannot be used to copy to or from a hard disk. The /l switch instructs DISKCOPY to copy only the first side of each disk, even if the disk is double-sided. The /v switch instructs DISKCOPY to perform a utility check on the data being copied.

Syntax:

DISKCOPY *drive1: drive2 [/l][/v]*

UNDELETE

UNDELETE is a DOS fault tolerance feature which allows recovery of deleted files.

Syntax:

UNDELETE *[drive:] [path] filename [/switches]*

When run in this manner, the undelete command requires you to supply the first letters of the files to be recovered. If the portion of the hard drive on which the file was stored is partially or entirely overwritten, you will probably not be able to recover the file.

Undelete has a sentry feature, activated by the command UNDELETE /SC. This is a memory resident program that creates a hidden directory in which deleted files are placed. With the sentry method, the undelete command does not require the user to input the first letters of files, and you do not have to worry about overwriting the portion of the disk on which the files to be recovered are stored.

SYS

This command is used to copy the system files from the hard drive or a floppy to another drive. SYS transfers the DOS system files (IO.SYS and MSDOS.SYS), plus the COMMAND.COM file, to a formatted disk without requiring reformatting.

Syntax:

SYS *[drive1:][path] drive2:*

FORMAT

Formatting Floppy Diskettes

A double density disk can be used in both low- and high- capacity disk drives, but a high-density disk can only be used in a high-capacity drive. DOS allows a high-capacity drive to format both double-density and high-density floppy disks. However, to do so, we must tell DOS exactly how we want the disk formatted. The command combinations below are the most commonly used.

- FORMAT A: – formats a 1.2 MB 5.25-inch disk in a 1.2 MB disk drive or a 1.44 MB 3.5-inch disk in a 1.44 MB disk drive
- FORMAT A: /4 – formats a 360 KB 5.25-inch disk in a 1.2 MB disk drive

- FORMAT A: /f:720 – formats a 720 KB 3.5-inch disk in a 1.44 MB disk drive
- FORMAT A: /S – formats a floppy disk and simultaneously transfers the system files

Formatting Hard Drives

The format command is also used to format hard drives. Before formatting a hard drive, make sure you have a full backup of that drive.

Syntax:

FORMAT *drive:*

UNFORMAT

UNFORMAT, like UNDELETE, is a DOS fault tolerance feature. It allows the recovery of files from an accidental disk format.

Syntax:

UNFORMAT *drive: [/switches]*

There are a few key points to remember about the UNFORMAT command: the results of this command vary from system to system; it cannot be used on a network drive; and there is a possibility that UNFORMAT may not be able to recover all the files it finds.

XCOPY

The XCOPY command allows you to copy files and directory trees. You use XCOPY command instead of COPY command when you want to copy all files and subdirectories from one location to another location.

Syntax:

XCOPY *source destination [/switches]*

Several switches can be used with the XCOPY command:

- /A – copies only those files which have their archive attribute set, without changing the archive attribute of the source files
- /M – copies only those files which have their archive attribute set, while turning off the archive attribute of the source files
- /D:*date* – copies only those files which were modified on or after the specified date
- /P – prompts to confirm the creation of each destination file
- /S – copies a directory and all its subdirectories, creating identical subdirectories on the target disk if they do not already exist, but not copying any empty source directories onto the target disk
- /E – copies empty source subdirectories onto the target disk; may only be used with /S
- /V – verifies the copied data
- /W – prompts to press a key before copying

TREE

This command displays the subdirectory structure of a directory. If you invoke this command without any parameters, the subdirectory structure of your default location is displayed. The /F switch includes the file names in each subdirectory.

Syntax:

TREE *[drive:][path][/switches]*

DELTREE

This command removes a directory, including all its files and subdirectories. This command has the ability to erase or overwrite your files. /Y prompts for confirmation of the delete.

Syntax:

DELTREE *[/switches][drive:][path] directoryname*

MEM

This utility is only available with DOS v5.0 and above. This command shows the available RAM and the total amount of installed RAM. You can use the /C (classify) switch with the MEM command to see the conventional, expanded, and extended memory, and the /P (pause) switch to display one screen at a time.

Syntax:

MEM *[/switches]*

CHKDSK

The CHKDSK utility tells the size of the specified drive, available space, lost clusters, lost sectors, and corrupted data. If CHKDSK is used with the /F switch, it fixes any low-level disk errors.

Syntax:

CHKDSK *[drive:][path][filename][/switches]*

SCANDISK

SCANDISK is a DOS utility that analyzes and repairs most logical and physical errors on a disk. It normally fixes errors on storage devices such as hard disks, floppy disks, RAM drives, and DBLSPACE compressed drives.

Syntax:

SCANDISK *[drive:][/switches]*

MSD

This command (short for Microsoft Diagnostics) is best run in DOS without Windows running. It provides detailed technical information about your computer. The information provided includes the model and processor of the computer, memory, video adapter, DOS version, mouse, other adapters, disk drives, LPT ports, COM ports, IRQ status, terminate and stay resident (TSR), and device drivers.

Syntax:

MSD [/switches]

PRINT

This utility is used to print text files to a LPT1 printer.

Syntax:

PRINT [drive:][path] filename [/switches]

SHARE

SHARE is used to make copies of files that are used by more than one program. One example of a program that uses multitasking is Microsoft Windows.

Syntax:

SHARE [/f:space][/l:locks]

MSBACKUP

The DOS MSBACKUP utility is used to make backup of floppy diskettes, and hard disk. Only the RESTORE utility can be used to retrieve the information saved by the MSBACKUP utility.

Syntax:

MSBACKUP *filename [/switches]*

FDISK

FDISK displays a summary of the partitions currently residing on the hard disk, and allows you to create new partitions and delete old partitions.

You can delete a partition using FDISK. When a partition is deleted all data on that partition is also deleted, and cannot be recovered later.

Other DOS Conventions

|MORE

The |MORE command is used to display one screen at a time, typically with the DIR, SORT, and TYPE commands. To use this command, the TEMP variable must be set (usually in the CONFIG.SYS file). TEMP specifies a directory that DOS can use to store temporary data.

DOSKEY

DOSKEY is an example of a TSR (Terminate and Stay Resident) program. Once you have invoked DOSKEY, it stays in memory until the system is restarted.

The DOSKEY command creates a small database that remembers the commands entered by the user. It occupies approximately 4 K of RAM. This command is very helpful if the user is creating a directory structure that requires long paths. The user can use the arrow keys on the keyboard to scroll through the contents of this database, and can execute any command in the database with or without modifying it. The F7 function key is used to see the contents of the database after invoking DOSKEY.

DOSKEY does not affect the standard DOS function keys, but it adds function keys of its own. The F7 key displays commands stored in memory, with their associated line numbers. <ALT F7> deletes all commands stored in memory. F8 searches memory for a command; type the first few characters of the desired command, then press F8. F9 prompts the user for a command line number, then displays the associated command for the number entered.

Chapter 12   DOS Directories and Files

Directory

For files or directories to be stored on a disk the disk must be formatted. The format process creates a special format (or structure) on the disk that allows files and directories to be stored there. During the formatting process a directory, called the Root, is created on the disk. Root is the first directory level, under which other directories and files can be stored. All disks used by DOS must be formatted and contain a root directory.

You can make three types of entries into a root directory: volume label (discussed earlier in this book), subdirectory, and file.

Subdirectory

A subdirectory is a directory that is created under a parent directory. You create a subdirectory to group all files of a similar type together. For example, you can create subdirectories under the parent directory, DOS, that represent different DOS versions. Under each subdirectory you would put files pertaining to that particular DOS version.

There are at least two reasons to use subdirectories. First, the operating system provides a limited number of entries in each directory. A hard disk's root directory allows up to 512 entries for files and directories. This capacity may be sufficient on a low-capacity hard disk, but a hard disk with many millions of bytes of storage may have more files than the directory permits. By using subdirectories, you can have a greater number of entries. Second, if you give the same file name in the same directory the new file overwrites the old one. By using subdirectories you can extend the entries to as many files and directories as needed and you can have files of the same name in different directories.

Files

Each diskette or hard disk is capable of storing many files, and each file must have a unique file name (within a directory or subdirectory). Files stored on separate directories or disks can have the same file name. As files are created and accessed, DOS keeps a record of each that includes the file name, extension, file size, date and time the file was last changed, file attributes, and sector where the file is located.

These conventions must be followed when naming a file:

Use a maximum of eight characters, then a period, and then an extension of up to three characters.

The following characters cannot be part of a file name:

" / \ [ ] : * < > | + = ; , ?

A file name cannot include a space.

These are some examples of valid file names:

POUND#
6-24-93
ROOM123
MYFILE

Even if the file name is entered in lowercase characters, DOS converts them into uppercase.

Special DOS Characters

DOS reserves the following characters:

- ? and * – the two DOS wildcard characters, discussed later
- . – separates a file name from its extension
- : – identifies drives and devices
- \ – identifies DOS paths
- / – identifies parameters
- < and > – data redirection symbols
- + – joins (adds) separate files into one file

File name Extensions

To distinguish between different files with the same name, you may add an extension of up to three characters to the file name. The extension must be separated from the file name by a period (.).

The following DOS extensions are used to denote a specific type of program or file:

- .BAK – Backup file. Some programs make a backup file whenever the file is updated.
- .BAS – File written in the BASIC programming language.
- .BAT – Batch file. Batch files can be edited.
- .C – File written in the C programming language.
- .COM – Command file. Command files are executed when typed at the DOS prompt.
- .DAT – Data file.
- .EXE – Executable program file. This kind of file contains instructions that are directly usable by the CPU.
- .OVL – Overlay file. This kind of file is called, as it is needed, into a reserved memory area by a program.
- .SYS – System file. A system file tells the computer what hardware is being used.

Wildcard Characters

Wildcard characters are used to tell DOS about a set of files with similar names or extensions. There are two wildcard characters that can be used with a variety of DOS commands:

- ? – represents one, and only one, character in a file name or an extension
- * – represents any number of characters (zero, one, or more than one) in a file name or an extension

For example, wildcard characters could be used when requesting the command processor to locate all the files with the .COM extension. To locate all the files with the .COM extension in the current directory, type the following:

DIR *.COM

Chapter 12: DOS Directories and Files

Type the following to delete all files in the current directory:

DEL *.*

To locate files that have some characters in common, both wildcards might be used. For example:

*.C??
FILE????.COM
????NAME.??M

Asterisk (*)

If you want to copy all the files on a floppy disk to a hard disk, type:

COPY A: *.* C:

The first asterisk is for all file names, and the second asterisk is for all extensions.

To copy all files with the extension .COM from a hard disk to a floppy disk, type:

COPY C: *.COM A:

Question Mark (?)

The question mark represents a single character, and can be used in a file name or an extension. Suppose you have a group of files which are different versions of the same file (ADV1, ADV2, ADV3, etc., all .DAT files). We can copy all of these files at once using the question mark.

Syntax:

COPY C: ADV?.DAT A:

The ? replaces the different version number in these files.

Batch Files

System processes can be automated through the use of batch files. Batch files are text files that contain a series of statements that are processed in order. These standard text files can be created and edited with any text editor. Standard DOS commands, batch processing commands, and other executable files can be included in DOS batch files. To start a batch file from the system prompt, the name of the file must be typed and the <ENTER> key must be pressed. To stop a batch file during execution, you can press <CTRL C> or <CTRL BREAK>.

Creating a Batch File

The DOS COPY CON command can be used to copy text typed from the keyboard console (con) and place the text into a batch file. DR DOS and MS DOS provide a text editor, invoked with the EDIT command, to create batch files. EDIT is a simple text editor that can be used to create a batch file. A batch file can be given any file name that is valid for any other type of file, but its extension must be .BAT.

Older versions of DOS have a text editor called EDLIN, but this is generally not used today.

The following commands are used only in batch files:

- REM
- ECHO
- PAUSE

REM

The REM command provides a way for inserting comments into a batch file. (REM stands for remark.) REM should appear at the beginning of each line that contains remarks. DOS ignores lines that begin with the REM command.

Syntax:

REM *remark*

Example:

REM The following command clears the screen.
CLS

ECHO

The ECHO command specifies whether the lines following it will be displayed as well as executed. ECHO OFF indicates that the commands following it will be executed but will not be displayed. With ECHO ON, the commands will be both executed and displayed. The ECHO command can be used in the AUTOEXEC.BAT file to generate a message that will be displayed every time the computer is turned on.

PAUSE

The PAUSE command suspends the execution of a batch file. Whenever a PAUSE command is used DOS displays the following message:

Press any key to continue...

DOS Configuration Files

There are two files that are executed automatically to configure your computer whenever you boot the computer:

- CONFIG.SYS
- AUTOEXEC.BAT

The CONFIG.SYS File

This file loads device drivers, changes default DOS settings, sets system environmental information, and specifies system parameters. CONFIG.SYS is a standard text file that can be created with COPY CON or any text editor (such as EDLIN or EDIT). The CONFIG.SYS file may look something like this:

```
FILES=40
BUFFERS=20
DEVICE=C:\DOS\MOUSE.SYS
```

The AUTOEXEC.BAT File

This file contains DOS commands that will execute when the system starts up or is reset. The AUTOEXEC.BAT file must reside in the root directory of the startup drive. If an

AUTOEXEC.BAT file does not exist, the system will prompt for the date and time at start up. AUTOEXEC.BAT is a standard text file, created by a text editor (such as EDLIN or EDIT). AUTOEXEC.BAT files commonly contain the PROMPT command and the commands that set the default drive and directory.

These two files will be discussed in detail in Chapter 14 of this book.

Chapter 13   Disks and Drives

Hard Disks and Floppy Disks

A computer uses two types of memory to store data and programs (information). One type is called Random Access Memory (RAM), or volatile memory. While the computer is processing information, it uses RAM to temporarily store the information. Information stored in RAM is lost if you turn off or reboot the computer. The other type of memory is called auxiliary memory. It is used to store information permanently; this information is not lost when you turn off or reboot the computer. The most common examples of auxiliary memory are hard disks and floppy disks.

Floppy Disks

Floppy disks (the terms "floppy disk" and "diskette" are used interchangeably) are widely used for computer data storage. The two most common sizes for floppy disks are 5.25" and 3.5", although all modern PCs now have only 3.5" floppy drives. Floppy disks (and hard disks) must be formatted before storing data on them. When you format a disk, the operating system prepares the disk so it can store the data. Formatting checks the disk surface for damage and imperfections, divides and organizes the disk into electromagnetic divisions called tracks and sectors, and creates a File Allocation Table (FAT) to store the location of files on the disk. Files are stored track by track and sector by sector. A set of contiguous sectors is called a cluster. Files are not always stored in contiguous sectors. A disk is considered fragmented when files are not stored in contiguous sectors.

Disk Capacity and Density

The following table shows the relationship between floppy disk size, capacity, and density.

Size	Capacity	Density/Sides
5.25"	360 KB	double-density, double-sided
5.25"	1.2 MB	high-density, double-sided
3.5"	720 KB	double-density, double-sided
3.5"	1.44 MB	high-density, double-sided

The capacity of a disk is measured in either kilobytes (thousands of bytes) or megabytes (millions of bytes). A 5.25-inch high capacity disk drive can read and write both 1.2 MB disks and 360 KB disks, and a 3.5-inch high capacity disk drive can read and write both 1.44 MB disks and 720 KB disks. But a 5.25-inch low capacity disk drive can only read and write 360 KB disks, and a 3.5-inch low capacity disk drive can only read and write 720 KB disks. A drive can read disks of lower capacity, but not higher.

The density of a disk refers to the amount of information that can be stored on the disk. Data stored on disks can be lost if the disk is physically damaged or exposed to magnetic fields.

Hard Disks

Hard disks provide larger and faster auxiliary storage capabilities for personal computers. Hard disks consist of one or more rigid metal platters coated with a metal oxide material that allows data to be magnetically recorded on the surface of the platters. Each platter has two read/write heads, one above and one below. The hard disk is much faster than a floppy disk.

Like floppy disks, data on a hard disk is organized into tracks and sectors. Hard disks also use cylinders to organize data. A cylinder is made up of all the tracks on all the disk platters that lie in the same relative position. Cylinders are numbered according to the track they reference. For example, the outermost track is called track 0, and all the track 0s on all the platters are collectively referred to as cylinder 0. Hard disks are available in capacity sizes ranging from 10 MB to over 9 GB (9,000 MB).

Partitioning the Hard Disk

Creating logical areas in a hard disk is called partitioning. A hard disk must be partitioned before it can be formatted. Multiple logical partitions can be created on a single physical disk. To DOS, each partition appears to be a separate, logical, unique hard disk. During partitioning, an active logical partition must be identified. This is the partition that will be used to boot the system, so it is referred to as the bootable partition or the primary partition. After the primary partition has been defined, all other partitions are referred to as extended partitions.

In MS DOS 3.3, and all previous versions of DOS, you are limited to partitions of up to 32MB. Later DOS versions allow much larger partitions. If the DOS version restricts your partition to a size less than the capacity of your disk, you can create multiple partitions. Also, when you want to use two or more operating systems, you can create a separate partition where each operating system will store its own file system. The DOS FDISK program invokes a menu that is used to create logical partitions.

Formatting the Hard Disk

Each partition is formatted separately on the hard disk. The FORMAT command is used to format the hard disk. The first or active partition can be formatted with the "/s" switch to make it bootable by typing:

FORMAT C: /s

IBM compatible computers can be booted from the A: and C: drives, but not from the B: drive.

To format other partitions, type FORMAT and the specific drive.

Example:

FORMAT D:

After DOS has formatted the hard disk it will show the total number of bytes in the disk, the number of bytes available, and the number of bytes in bad sectors. It is normal to have 1% to 2% of the total bytes in bad sectors.

The drive letter designation displayed at the DOS prompt is known as the active or default drive. This is the drive that DOS automatically uses until you specify a different drive letter.

The default drive assignment will vary depending upon the specific hardware you are using. A two-diskette system typically assigns drive A: as the default drive. If there is only one floppy disk drive, DOS will behave as though there were two, and letters A: and B: will both reference the same disk drive. The first hard drive is usually referenced by C:. Any additional drives (hard, CD-ROM, or floppy) then become drive D: and so on.

Note: Formatting a disk or changing the partition size of a disk will destroy all data on the disk or on the partition. The UNFORMAT command can be used to restore a disk that was erased by the FORMAT command. UNFORMAT restores only local hard drives and diskettes. It cannot be used on network drives.

Chapter 14 Memory Management

Memory Optimization

System optimization refers to the well being of the computer. The system configuration and its use of memory are of primary importance in tuning the performance of the PC. The system configuration files, CONFIG.SYS and AUTOEXEC.BAT can be edited using the EDIT command.

There are two ways to optimize these files. The first is to let the computer try its hand at optimization using the MEMMAKER command. The computer will survey its configuration and will optimize memory by moving device drivers and memory resident programs to upper memory.

MEMMAKER should not be used while Windows is running.

Although MEMMAKER is helpful in obtaining more conventional memory, you can also optimize your memory manually. This approach can sometimes yield greater results.

F5 and F8 Function Keys

These two function keys are used to bypass configuration files during the bootup process. F5 is used to directly bypass both CONFIG.SYS and AUTOEXEC.BAT. F8 is used to step through

CONFIG.SYS and AUTOEXEC.BAT files one line at a time, giving the user the option to execute or not execute each line. This is useful in troubleshooting configuration file problems to determine which line is causing a problem.

CONFIG.SYS

The CONFIG.SYS file is an ASCII text file that contains instructions for DOS regarding system configuration. It resides in the root directory, and is read by DOS only once, at startup. Since it is an ASCII file, it can be easily edited using any text editor or word processor that saves files in ASCII format. You must reboot your computer to activate any changes you make to this file; you cannot run this file manually.

CONFIG.SYS loads device drivers, changes default DOS settings, sets system environmental information, and specifies system parameters.

While changing an existing CONFIG.SYS file, use the EDIT utility (instead of COPY CON) so that you can use the exact syntax of the device drivers without retyping.

As you make changes in a CONFIG.SYS file to optimize memory, you can reboot after one or more changes and use MEM /C /P to check the effect of the changes on conventional and other memory categories.

Before you start modifying existing configuration files, make a separate copy of the existing configuration file(s) under a different name so that if something goes wrong you can return the computer to its prior configuration.

The order of commands in the CONFIG.SYS file is very important. For example, if you intend to use HIMEM.SYS, it should precede the commands for loading drivers or other software that uses HIMEM or upper memory.

Some of the commands commonly found in this file are:

- BREAK
- BUFFERS
- DEVICE
- DEVICEHIGH
- DOS
- FCBS
- FILES
- INSTALL
- LASTDRIVE
- REM
- SHELL
- STACKS

BREAK

This command tells DOS to cancel any function if the user presses <CTRL BREAK>. The default status of BREAK is OFF.

Syntax:

BREAK=ON/OFF

Example:

BREAK=ON

This example causes DOS to watch for a cancellation command during any DOS function call. The default status for BREAK is OFF.

BUFFERS

This command tells DOS how much memory to set aside for disk buffers. A buffer is an area in the Random Access Memory (RAM) set aside for temporary data storage. When DOS reads a file, 512 bytes of information is stored in each buffer, making it accessible to the CPU. If there are not enough buffers, the system may slow down or the application may not work. Too many buffers will also slow down your system. Set your buffers to the highest number needed for any application you are running. You can specify any number of buffers from 1 to 99. (The default is 15 for most systems.) Each buffer uses 528 bytes of conventional memory.

Syntax:

BUFFERS=*n[,m]*

where *n* specifies the number of disk buffers and *m* specifies the number of buffers in the secondary buffer cache (read-ahead buffers). The value of *m* can be any number from 0 to 8. Default is 0.

Example:

BUFFERS=30,8

The following is a table of the suggested use of this command in the CONFIG.SYS file.

Hard Disk Size (in MB)	Suggested Number of Buffers
20-32	20
33-80	30
81-120	40
Above 120	50

DEVICE

This command installs device drivers. Every hardware device on a system uses a software routine (called a driver) that tells DOS how to properly interface with the device. When a new device is added for which neither DOS nor the ROM BIOS already has a driver, a device driver needs to be defined in the CONFIG.SYS program.

Syntax:

DEVICE=*drive:\path\filename /switches*

Example:

DEVICE=C:\MOUSE\MOUSE.SYS

This example loads the device driver named MOUSE.SYS into memory from the MOUSE directory of the C: drive.

Some of the most commonly used device drivers are as follows:

- ANSI.SYS – extended screen and keyboard device driver
- DRIVER.SYS – assigns logical drive letter to floppy disk drives
- EMM386.EXE –installs expanded and reserved memory support for computer systems with extended memory

- **HIMEM.SYS** – loads extended memory support using Microsoft's XMS extended memory specification
- MOUSE.SYS – allows a mouse to be used with the computer
- POWER.EXE – lowers the rate of power consumption when devices are idle
- RAMDRIVE.EXE – initializes a RAM disk
- SETVER.EXE – installs a list of software applications that require DOS to supply an earlier version number
- SMARTDRV.EXE – installs double-buffering for a SMARTDrive cache
- VDISK.SYS – allows the creation of a virtual disk (RAM disk)

DEVICEHIGH

This command installs device drivers in reserved memory, if space is available. If you include this command in your CONFIG.SYS file, you must first use HIMEM.SYS and EMM386.EXE.

Syntax:

DEVICEHIGH /*switches=drive:\path\filename*

Example:

DEVICEHIGH=C:\DOS\SETVER.EXE

This example loads the program named SETVER.EXE in the upper memory.

DOS

This command loads the DOS operating system in conventional, extended, or reserved memory, depending on the switches used. Use this command after HIMEM.SYS and EMM386.EXE. The switches used with this command are as follows:

- high – loads DOS in extended memory; cannot be used with low
- low – loads DOS in conventional memory; cannot be used with high
- UMB – loads DOS in reserved memory; cannot be used with NOUMB
- NOUMB – forces DOS to ignore upper memory blocks; cannot be used with UMB

You must use at least one switch with the DOS command.

Syntax:

DOS=high/low, umb/noumb

Example:

DOS=HIGH, UMB

This example loads DOS into reserved memory, with any remainder placed in extended memory.

FCBS

This command, mainly used in networking schemes, specifies the number of open files using the file control block. The *maximum* parameter in the syntax shown below indicates the maximum number of open file control blocks. This can be any number from 1 to 255, with a default value of 4. The *open* parameter indicates the number of files that will not automatically close if processing attempts to open more files than are allowed by the *maximum* parameter. Use this command only if the program requires it. Most newer programs do not require file control blocks.

Syntax:

FCBS=*maximum,open*

Example:

FCBS=20,4

This example specifies a maximum of 20 open file control blocks, with up to 4 files protected from automatic closing if processing attempts to open more than 20 files.

FILES

This command is required for systems that have multitasking capability. Some operating systems, including Windows, allow you to run multiple programs simultaneously. This command sets the maximum number of files that can be open simultaneously.

Syntax:

FILES=*n*

where *n* represents the maximum number of open files. The maximum value of *n* is 255, with a default value of 8. If you exceed the maximum number of open files during processing, DOS displays the message "Too many files are open."

Example:

FILES=30

This example specifies that a maximum of 30 files may be open at one time.

INSTALL

This command loads terminate-and-stay-resident (TSR) software. The INSTALL command helps to reduce conflicts by loading such programs in areas of memory where they are least likely to cause RAM addressing conflicts with other applications.

Syntax:

INSTALL=*drive:path filename*

Example:

INSTALL=C:\DOS\SHARE.EXE

This example installs SHARE.EXE from the DOS directory on the C: drive.

LASTDRIVE

The LASTDRIVE command configures DOS to reserve a certain number of drives as local disk drives. DOS can recognize up to 26 disk drives. If a network workstation has more than five logical disk drives, setting the LASTDRIVE command ensures that DOS will recognize all the local drives. The *drive* parameter can be any letter from A to Z. In a network environment, this letter should always be Z.

Syntax:

LASTDRIVE=*drive*

Example:

LASTDRIVE=Z

This example sets a maximum of 26 logical drive letters. Each drive letter after the first five (any letter beyond E) uses 81 bytes of RAM.

[handwritten: 1,701 Bytes beyond E]

REM

This command indicates a comment line that is ignored by DOS in processing.

Syntax:

REM *comment*

Example:

REM This is just a comment line.

SHELL

This command installs an alternate COMMAND.COM file, and also changes the size of the environment space. The default size of the environment space is 160 bytes; the maximum size is 32,768 bytes. Two switches are used with this command:

- /P – loads the specified COMMAND.COM file as the primary processor (in the absence of this parameter, the specified COMMAND.COM file is loaded as the secondary processor)
- /E:# – indicates the size of the environment space, where # represents the size of the environment space in bytes

Syntax:

SHELL=*drive:\path* COMMAND.COM *[/switches]*

Example:

SHELL=C:\COMMAND.COM /E:256 /P

This example loads the copy of COMMAND.COM on drive C: as the primary processor, and changes the size of the environment space to 256 bytes.

STACKS

This command sets dynamic allocation of stack space. Dynamic stack space allocation permits multiple interrupt calls to call each other without crashing the system. The *frames* parameter sets the number of stack frames, which can be any number from 8 to 64. The default is 9 for most systems. The *size* parameter indicates the size of each frame; this can be anywhere from 32 to 512 bytes. The default is 128 for most systems. Do not use this command unless you see the message "Stack Overflow" or the message "Exception error 12".

Syntax:

STACKS=*frames,size*

Example:

STACKS=10,256

This example sets the dynamic stack capacity to 10 frames of 256 bytes each.

Sample CONFIG.SYS File

The following is an example of a simple CONFIG.SYS file.

```
DEVICE=C:\DOS\SETVER.EXE
DEVICE=C:\DOS\HIMEM.SYS
DEVICE=C:\DOS\EMM386.EXE NOEMS
DOS=HIGH, UMB
DEVICEHIGH=C:\MOUSE\MOUSE.SYS
INSTALL\C:\DOS\SHARE.EXE
BUFFERS=20
FILES=40
BREAK=ON
LASTDRIVE=Z
```

AUTOEXEC.BAT

Like CONFIG.SYS, AUTOEXEC.BAT is read when the computer boots up. This file, which is created by the user, creates the working environment for the user.

Some of the commands commonly found in this file are:

- @

- CALL
- ECHO
- MSCDEX.EXE
- PAUSE
- PATH
- PROMPT
- REM

@

This character is used with a command to suppress the display of the line on the screen when the command is run.

Example:

@ECHO Do not show this command.

CALL

This command is used in a batch file to call and execute another batch file. When using this command, you must specify the complete path of the batch file to be executed. For example, if you want to run a batch file named STARTNET.BAT (a file used to connect to a NetWare server) from the NWCLIENT directory that resides under the root directory of the C: drive, then you should type

CALL C:\NWCLIENT\STARTNET.BAT

ECHO

This command specifies whether to suppress or display batch file lines on the screen at the time of execution. ECHO ON displays your AUTOEXEC.BAT; ECHO OFF suppresses the display. ECHO OFF is normally used with @, as follows:

@ ECHO OFF

In this example, ECHO OFF suppresses the display of lines in your batch file, while @ suppresses the display of the line containing ECHO OFF. Therefore, nothing from the batch file will be displayed.

PAUSE

This command pauses the processing of a batch file until the user presses a key. It is commonly used to display messages to the user. When the PAUSE command is invoked, it displays the following message:

Press any key to continue…

When the user presses a key, the batch file resumes executing.

PATH

The PATH command creates a list of directories to search for the executables files of the operating system. Whenever a user uses any command, DOS searches for that command in all directories listed in the PATH command. This command is used in batch files with the following syntax:

PATH C:\DOS; C:\WINDOWS; C:\NWCLIENT

PROMPT

As explained earlier in this book, this command creates the system prompt. The prompt C:\> is commonly used; this prompt is created using the switches pg, as follows:

PROMPT pg

REM

This command makes any statement a remark statement. The system will read the line, but will not try to execute that line. REM is commonly used for troubleshooting configuration files or for documenting a file.

Example:

REM This is a remark statement.

Memory Optimization Tip

To optimize system memory through the AUTOEXEC.BAT file, we normally use the LH (LOADHIGH) command. This command moves the TSR (terminate-and-stay-resident) programs into Upper Memory Blocks (UMB), depending on the availability of space. This command can only be used if UMB is made available by loading the required drivers in the CONFIG.SYS file. For example, using this command, MSCDEX.EXE in the AUTOEXEC.BAT file can be loaded into upper memory, thus saving approximately 35 KB of conventional memory.

MSCDEX.EXE

If you have a CD-ROM drive, the AUTOEXEC.BAT file would contain a MSCDEX.EXE command. This command tells the

Introduction to Personal Computers for Networking Professionals

system to establish a drive letter for a CD-ROM drive, which allows the CD-ROM drive to function. The switches after the command load the device driver, which must match the name of the device driver for the CD-ROM drive in the CONFIG.SYS file.

Multiple Configuration

DOS 6.0 and later versions allow a user to create multiple configurations (startup menus) at startup time using the CONFIG.SYS and AUTOEXEC.BAT files. We will briefly go over the different commands used to create a very simple startup menu with the following options:

MS-DOS 6.22 Startup Menu

    Start DOS
    Windows 3.1
    Connect to NetWare Server

To create the above menu, we would use the following CONFIG.SYS and AUTOEXEC.BAT files:

CONFIG.SYS

```
DEVICE=C:\DOS\HIMEM.SYS
DEVICE=C:\DOS\EMM386.EXE NOEMS
DOS=HIGH, UMB
[MENU]
MENUCOLOR=15,1
MENUITEM=DOS, Start DOS
MENUITEM=WINDOWS, Windows 3.1
MENUITEM=NETWORK, Connect to NetWare Server
MENUDEFAULT=DOS, 15
```

Chapter 14: Memory Management

```
[DOS]
FILES=15
[WINDOWS]
FILES=40
[NETWORK]
LASTDRIVE=Z
[COMMON]
DEVICEHIGH=C:\DOS\SETVER.EXE
DEVICEHIGH=C:\MTM\MTMCDAI.SYS /D:MEMIDE01
```

AUTOEXEC.BAT

```
LH C:\DOS\MSCDEX.EXE /D:MEMIDE01 /M:10     -? Creates Disk Buffers
C:\DOS\SMARTDRV.EXE
@ECHO OFF
PROMPT $p$g
PATH=C:\DOS; C:\WINDOWS; C:\NWCLIENT
SET TEMP=C:\DOS
:DOS
:WINDOWS
C:\WINDOWS\WIN  .com
:NETWORK
C:\NWCLIENT\STARTNET.BAT
:END
```

Menu Blocks

To create multiple configurations, we start with menu blocks. A menu block is a series of commands in the CONFIG.SYS file that describes a simple menu of options to be displayed at startup. The menu block has to be enclosed in square brackets, as follows (in this case, the brackets are part of the syntax):

[MENU]

MENUITEM

Under the menu block, all the options are listed using the command keyword MENUITEM. For example:

MENUITEM=DOS, Start DOS

means that DOS is the title of the configuration block and Start DOS (an optional parameter) is the text to be displayed on the screen.

MENUCOLOR

The command MENUCOLOR sets the color of the menu.

Syntax:

MENUCOLOR=n,m

The first parameter, n (15 in our sample file) defines the text color, whereas the second parameter, m (1 in our sample file) defines the background color. Below is a summary of color codes used in DOS.

Code	Color
0	Black
1	Blue
2	Green
3	Cyan
4	Red
5	Magenta
6	Brown
7	White
8	Gray

9	Bright blue
10	Bright green
11	Bright cyan
12	Bright red
13	Bright magenta
14	Yellow
15	Bright white

MENUDEFAULT

This sets the default option of the user.

Syntax:

MENUDEFAULT=*configurationblock,n*

The *n* parameter represents the number of seconds that the system will wait for the user to choose an option before the system executes the default option. In our example, if the user did not choose another option, DOS would start after 15 seconds.

Configuration Blocks

The configuration blocks [DOS], [WINDOWS], and [NETWORK] are used to configure the individual menu options. This is a good way to optimize system memory. Commands that are to be used with all of the menu options are always listed under the [COMMON] configuration blocks. This way, you don't have to write the same commands over and over for each option. These configuration blocks are entered in square brackets, just as the menu block is.

In the AUTOEXEC.BAT file, we use the corresponding section blocks to execute the menu options or to further enhance their functionality. For example, :DOS, :WINDOWS, :NETWORK, etc.

Chapter 15   Emergency Preparedness

Bootup Order/Procedure

The following is the typical bootup order:

- POST (Power On Self-Test)
- BIOS
- CMOS
- CONFIG.SYS
- COMMAND.COM
- AUTOEXEC.BAT

You can use F5 and F8 to override potentially troublesome configuration files.

BIOS Setup

Each computer stores its own setup information. This is accessed at bootup. The setup file is used to troubleshoot and change configuration including hard drives, boot drive order, and many other features. How the setup file can be changed depends on the make of the computer and the manufacturer of the BIOS. The following table provides a general guideline:

BIOS Manufacturer	Keys to access setup
AMI	<DELETE>
Award	<CTRL ALT ESC>
Phoenix	<CTRL ALT ESC> or <CTRL ALT INS> or F10
Dell	<CTRL ALT ENTER>

Compaq frequently requires you to use a separate system disk or CD to access the system setup file.

Prevention of problems is significantly more efficient in terms of cost and time than trying to correct problems after the fact. Several suggestions follow which could help save time in the event of a disk crash or some other disastrous event.

Write or print out your CMOS settings. When you need to see them, it may be too late to access them on the computer—you may need a hard copy.

Create an emergency disk—a bootable floppy with a current copy of CONFIG.SYS, AUTOEXEC.BAT, and all programs and files called by those two files. In addition, keep a copy of FDISK.EXE, FORMAT.COM, SYS.COM, SCANDISK.EXE, SCANDISK.INI, UNDELETE.EXE, UNFORMAT.COM, EDIT.COM, and QBASIC.EXE. If you are running Windows 3.1, you should also copy all .INI and .GRP files, especially SYSTEM.INI and WIN.INI. If you are running Windows 95, make a startup disk.

Write or print out your IRQ and DMA settings.

Use virus protection software.

Use surge protectors.

Use a UPS (Uninterruptible Power Supply) for important PCs.

Maintain a backup system with a regular schedule and convenient off-site storage. Two backups are ideal—one on-site and one off-site.

Unit 2 Review

Matching

Match each command to its description.

Q1. Makes identical floppy diskettes. _____
Q2. Removes an empty directory. _____
Q3. Displays directory path. _____
Q4. Copies all files. _____
Q5. Makes a directory. _____
Q6. Displays a directory listing of WP. _____
Q7. Erases all data. _____
Q8. Changes directory. _____
Q9. Changes the attributes of a file. _____
Q10. Renames a file. _____
Q11. Displays directory listings by page. _____
Q12. Used to partition your fixed drive. _____

a) ATTRIB   g) RD
b) FDISK   h) COPY *.*
c) FORMAT   i) DIR WP
d) DISKCOPY   j) CD
e) REN   k) DIR /P
f) PATH   l) MKDIR

Chapter 15: Emergency Preparedness

True or False

Mark each statement as true or false.

_____ Q1. COPY can be used to add the contents of two files.

_____ Q2. COMMAND.COM is a hidden file.

_____ Q3. Hyphens are used in the TIME command to separate minutes and seconds.

_____ Q4. In the PROMPT command, $n represents the default drive.

_____ Q5. RMDIR and MKDIR perform the same function.

_____ Q6. When used with the DIR command, /W causes DOS to present files in five columns across the screen.

_____ Q7. The CLS command clears the screen as well as the prompt.

_____ Q8. Using the TYPE command on a file with a .COM or .EXE extension produces a series of readable symbols.

_____ Q9. When using the TIME command, the time can only be entered in 24-hour format.

_____ Q10. The DEL command can always be used to delete any directory, no matter what is in it.

Multiple Choice

Select the best answer.

Q.1. How does DIR /O display directories and files?

    a) In alphabetical order
    b) In chronological order
    c) In order of file size
    d) None of the above

Q.2. What will be copied of you type COPY *.*?

    a) All files
    b) No files
    c) Files with any extension
    d) System files

Q.3. What should you type with the PROMPT command to create an equal sign (=)?

    a) $b
    b) $g
    c) $e
    d) $q

Q.4. How many switches does the DEL/ERASE command have?

   a) Two
   b) Three
   c) One
   d) None of the above

Q.5. What information is listed by the DIR command?

   a) Volume label
   b) Directory path
   c) File size
   d) Number of files and free space

Short Answer

Briefly explain what will happen when you enter each of the following commands, assuming that you have the directory structure indicated by the prompt given in each question.

Q1.   At the C:\PCAGE> prompt, type COPY *.??M C\SS

_____
_____
_____

Q2.   At the C:\> prompt, type CD\PCAGE\TEST\DBASE

_____
_____
_____

Q3. At the C:\> prompt, type PATH C:\; C:\DOS; C:\WINDOWS

Q4. At the C:\WINDOWS> prompt, type DIR/W/P/O

Q5. At the C:\DOS> prompt, type TYPE CONFIG.SYS

Q6. At the C:\WP> prompt, type DEL ??*.?*E

Q7. At the C:\> prompt, type COPY *.* C:

Q8. At the C:\> prompt, type PROMPT d_$t

Chapter 15: Emergency Preparedness

Q9. At the C:\DOS> prompt, type TIME 10

_____
_____
_____

Q10. At the C:\> prompt, type COPY FIRST C:\WINDOWS

_____
_____
_____

Hands-On Exercise: Directory Management

Perform the following steps.

1. Make the following directories and subdirectories in the root directory.

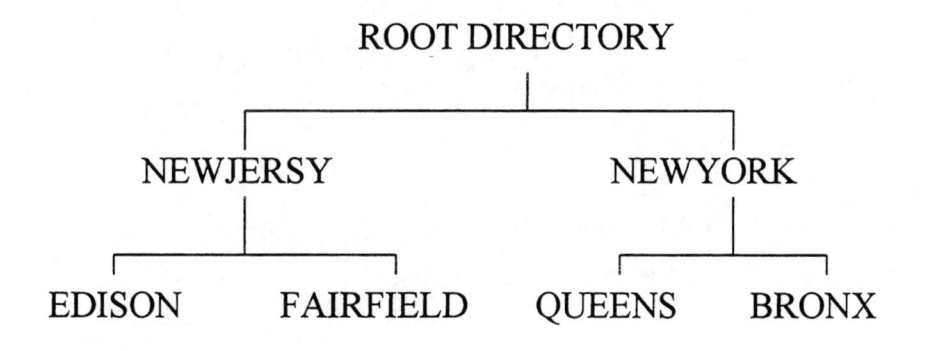

```
                    ROOT DIRECTORY
                          |
            ┌─────────────┴─────────────┐
         NEWJERSY                    NEWYORK
            |                            |
       ┌────┴────┐                  ┌────┴────┐
    EDISON   FAIRFIELD           QUEENS    BRONX
```

2. Copy all files with an extension of .EXE from DOS to EDISON.

© 1998 · PC Age, Inc. All Rights Reserved · 20 Audrey Place · Fairfield, NJ 07004 · U.S.A. · Tel: 973-882-5002
www.pcage.com

3. Copy all files with an extension of .SYS from DOS to QUEENS.
4. Copy all files with an extension of .COM from DOS to BRONX.
5. Verify the contents of each directory without leaving the root directory.
6. Using the COPY command, make a combined file of CONFIG.SYS and AUTOEXEC.BAT in FAIRFIELD, under the file name of AUTOCON.BAT.
7. Verify the result of Step 6.
8. Delete the EDISON directory.
9. Using the XCOPY command, copy the complete directory structure of the NEWYORK directory under the NEWJERSY directory, using the directory name DENVER.
10. Delete the NEWJERSY directory.
11. From the root directory, using one command, go to the BRONX directory.
12. Change directory to QUEENS.
13. Change the prompt to your own name.
14. Check the system date, time, and version.
15. Change the name of CONFIG.SYS in QUEENS to CONFIG.COM in BRONX.
16. Change the prompt back to the way it was before starting this exercise.
17. Rename the BRONX directory to MANHATTN.
18. Remove all the remaining directories.
19. Clear the screen.

Hands-On Exercise: Memory Management

Load the following commands and drivers in such a way that when you are finished, you have approximately 600 KB as free conventional memory in your system.

Note: Before beginning this exercise, save your current CONFIG.SYS file as CONFIG.OLD, and your current AUTOEXEC.BAT file as AUTOEXEC.OLD. When you are finished with the exercise, restore these original files.

1. SETVER.EXE
2. FILES=40
3. BUFFERS=15,0
4. LASTDRIVE=Z
5. Driver for CD-ROM
6. SMARTDRV.EXE
7. PROMPT command
8. PATH command

[Handwritten annotations:]

Config.sys — items 1–4

Both ? Where — item 5 for CD-ROM

Autoexec.BAT — items 6–8

```
COPY AUTOEXEC.BAT to AUTOEXEC.OLD
COPY CONFIG.SYS to CONFIG.OLD

DEVICE = C:\DOS\HIGHMEM.SYS
DEVICE = C:\DOS\EMM386.EXE
DOS = HIGH, UMB
DEVICEHIGH = C:\DOS\SETVER.EXE
Files = 40
Buffer = 15,0
LASTDrive = Z
DEVICEHIGH = C:\MTM\MTMCDAI.SYS
DEVICEHIGH = C:\DOS\SMARTDRV.EXE
```

AUTOEXEC.BAT:
```
PROMPT $P$g
PATH C:\DOS
```

Unit 3: Windows

Chapter 16   Windows 95

Unlike prior versions of Windows, Windows 95 is not just an operating environment but also an operating system. It comes with extensive help programs and tutorials.

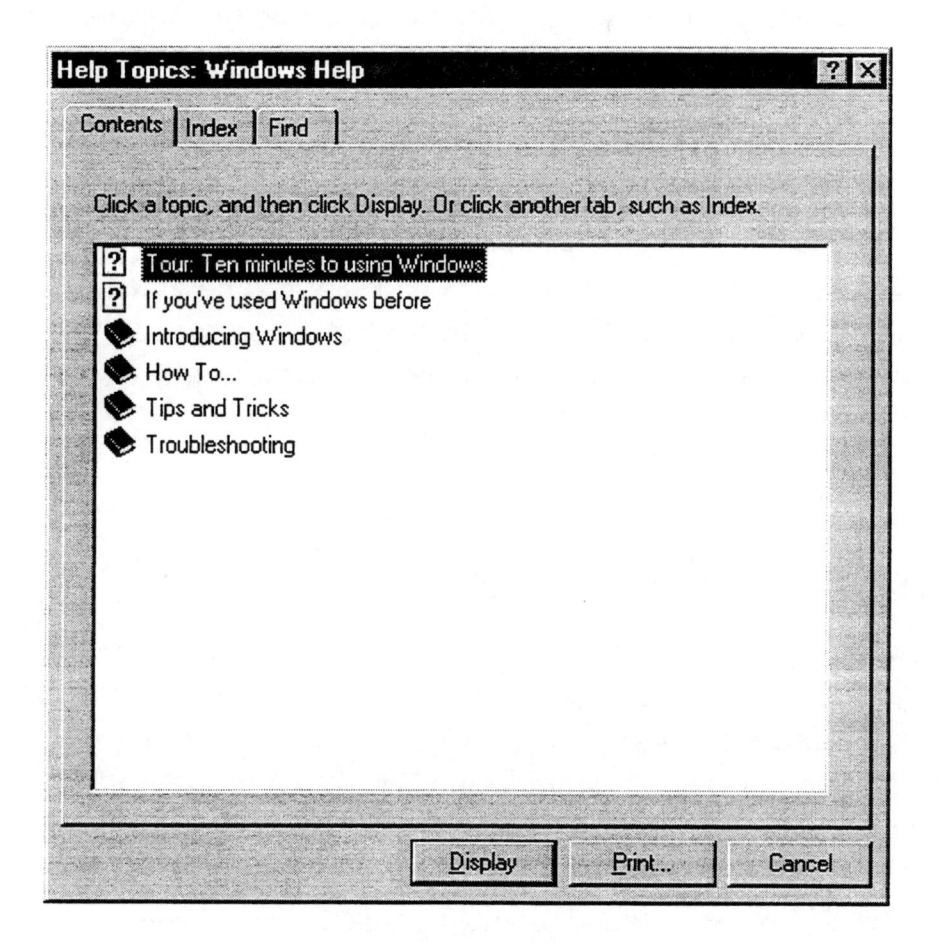

Figure 16-1 Help Screen

The help provided with Windows 95 is equally beneficial for advanced and novice users. To access this help, click on the Start button (located in the lower left-hand corner of the screen) and select Help.

New Features

Windows 95 offers an improved interface. Using the Start button you can quickly open programs, find documents, and use system tools. You can also use the task bar, at the bottom of the screen, to switch between different programs easily.

Windows 95 supports long filenames (of more than eight characters) and comes with Windows Explorer, a powerful tool for browsing and managing your files and folders. There is also good news for users who like games. In Windows 95, you will enjoy faster video capability for games, enhanced support for MS-DOS based games, and improved performance for playing video and sound files.

Installing Windows 95

Windows 95 has some minimum system requirements. It requires a 80386DX or higher processor with a minimum of 4MB of RAM. It requires a free disk space of 35 to 70 MB, depending on the setup options (Typical, Custom, Compact, etc.). If you install all the components, then the free space requirement will change.

Windows 95 can be installed as an Upgrade of any existing version of Windows, or it can be installed as a full version. The software is usually shipped on a CD, but is also available on 3.25" floppy disks.

Running Windows 95

Once you start Windows 95, the first screen you will see is the desktop.

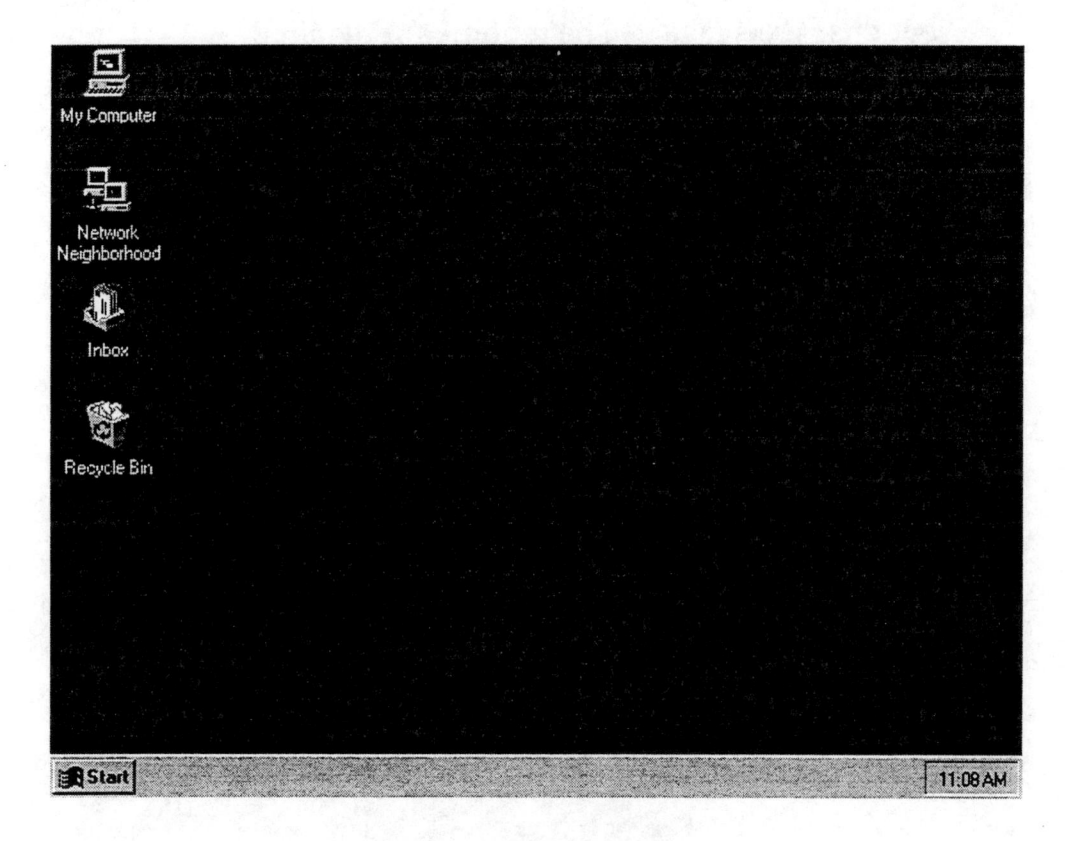

Figure 16-2 Desktop

You will see different icons on your desktop, depending on how your computer is set up. Some of the important icons are:

- My Computer – allows you to see what is inside your computer
- Network Neighborhood – allows you to view available network resources

- Recycle Bin – acts as a temporary storage place for deleted files. You have an option to recover these files, or to permanently delete them from the system. Windows 95 reserves 10% of your hard disk space for the Recycle Bin, but this percentage can be modified if required.

Using Windows 95

When you click the Start button, a menu pops up that contains everything you need to start using Windows.

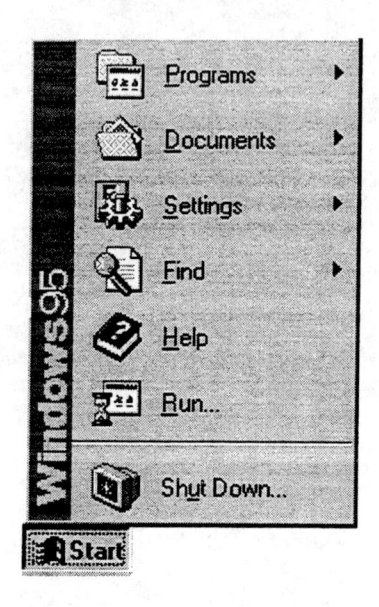

Figure 16-2 Start Menu

This menu contains the following commands:

Menu Item	Function
Programs	Displays a list of programs
Documents	Displays a list of recently opened documents

Settings	Displays a list of system components for which the settings can be changed
Find	Enables you to search for files, folders, shared computers, etc.
Help	Windows 95 online help
Run	Starts a program or opens a folder
Shut Down	Shuts down or restarts your computer

Running Programs

To run a program under Windows 95, click the Start button and move the pointer to Programs. This opens up your program menu.

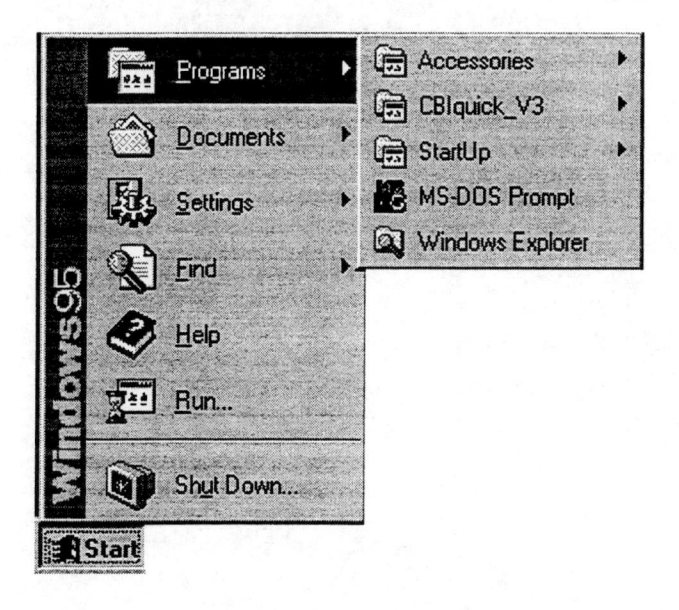

Figure 16-3 Programs Menu

If you have installed applications on your computer, you can run any application from the Programs menu by moving the pointer to that program and clicking it. It's that easy. You can also run a program from the Run option in the Start menu by selecting Run and typing in the name of the program (including the complete path, if necessary).

If you haven't installed any applications yet, you will only see the following options: Accessories, Startup, MS-DOS Prompt, and Windows Explorer.

Accessories

Accessories is a folder that contains the Multimedia and System Tools subfolders. It also contains some useful programs such as Calculator, Notepad, Paint, Phone Dialer, and WordPad.

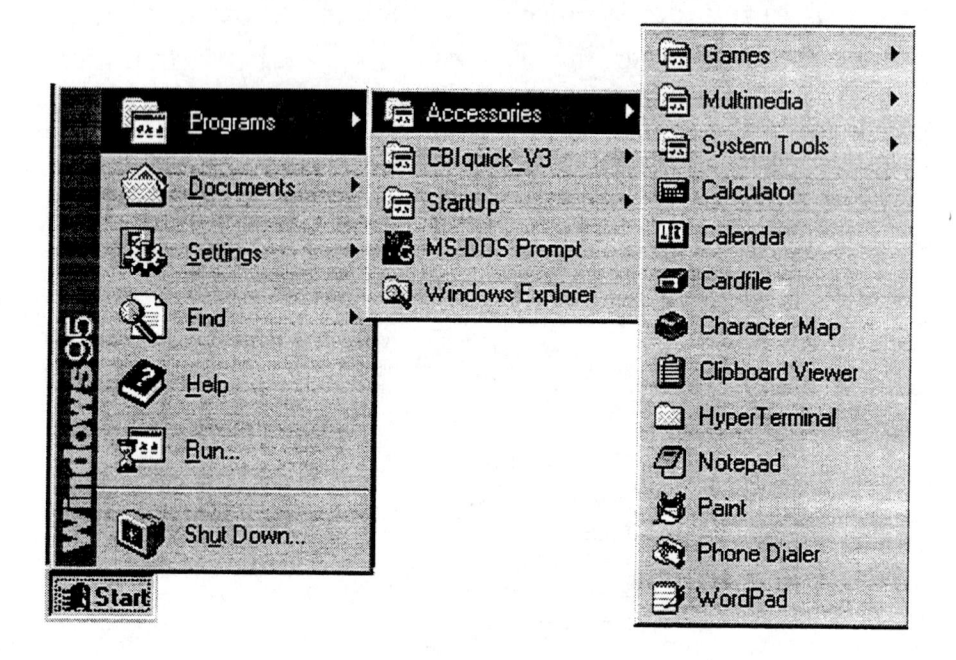

Figure 16-4 Accessories Menu

Startup

Startup is also a folder. It is used to configure programs to run automatically at Windows startup.

MS-DOS Prompt

This option allows you to temporarily switch to MS-DOS mode, from which you can run any MS-DOS based programs.

Windows Explorer

This tool is used to browse your computer's directory structure. It also helps you to create new folders, manage your files, and (if you are connected to a network) map and browse network drives.

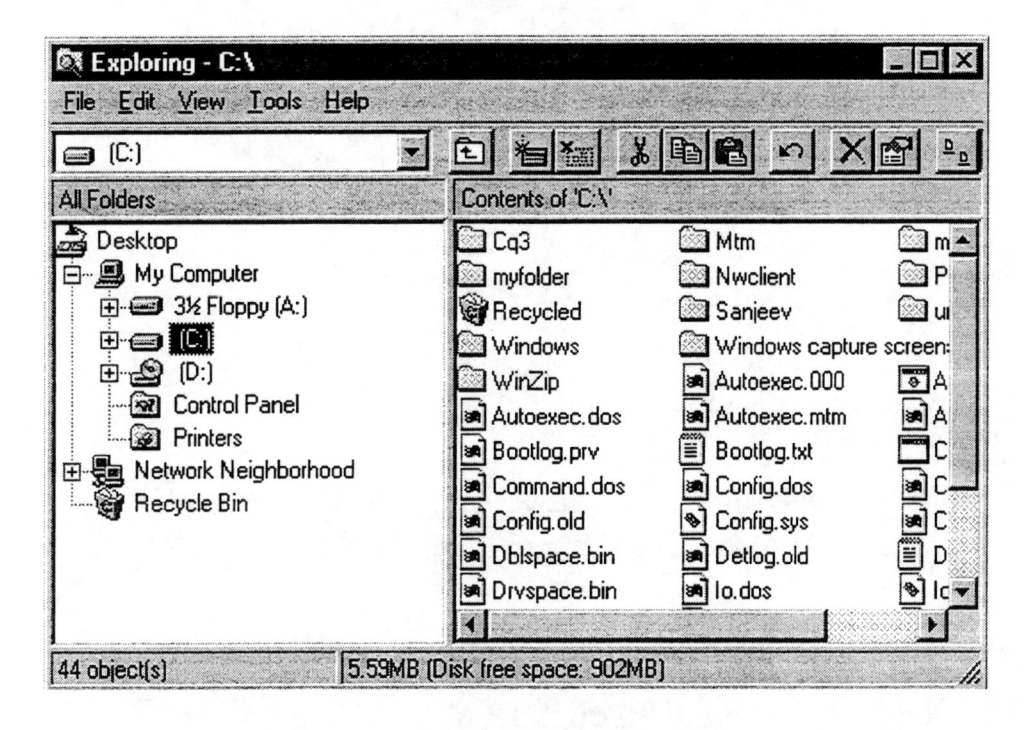

Figure 16-5 Windows Explorer

Shutting Down Your Computer

Windows 95 is a sensitive operating system. If it is shut down improperly—for example, by simply turning off your computer while Windows 95 is running—it can damage your system files.

To shut down Windows properly, click the Start button, then Shut Down, then the Yes button.

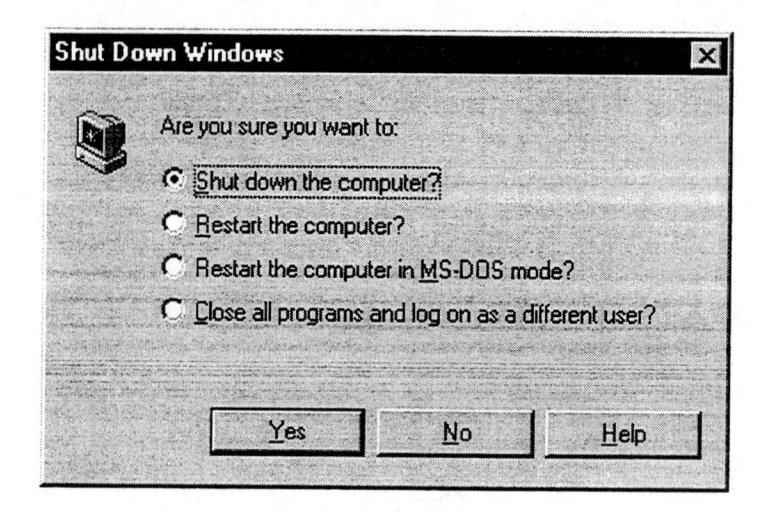

Figure 16-6 Shutdown Screen

Installing Software

There are two ways to install software on your system: with the Run command and with Add/Remove Programs.

Installing Software Using the Run Command

Click the Start button, then click Run. In the Run command line box, type *X*:\setup (where X represents the drive from which the software is being installed). Now follow the instructions on your screen. Some software uses INSTALL.EXE or INSTALL.BAT files instead of SETUP.EXE to invoke the installation process.

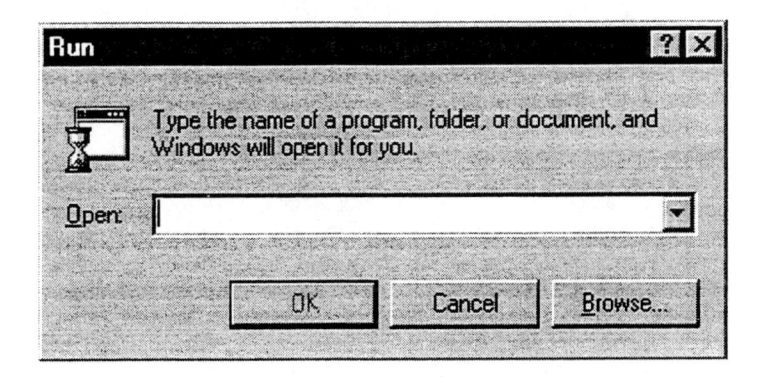

Figure 16-7 Run Command Line Box

Installing Software Using Add/Remove Programs

Click the Start button, point to Settings, then click Control Panel.

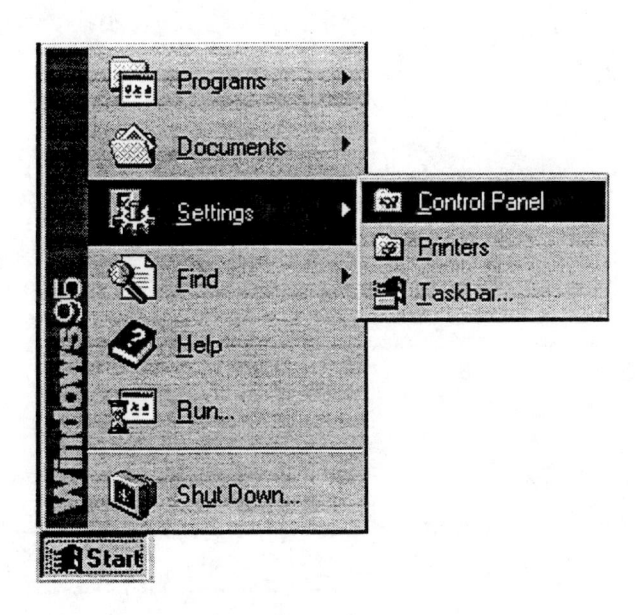

Figure 16-8 Selecting Control Panel

Double-click the Add/Remove Programs icon, then click the Install button. Now follow the instructions on your screen.

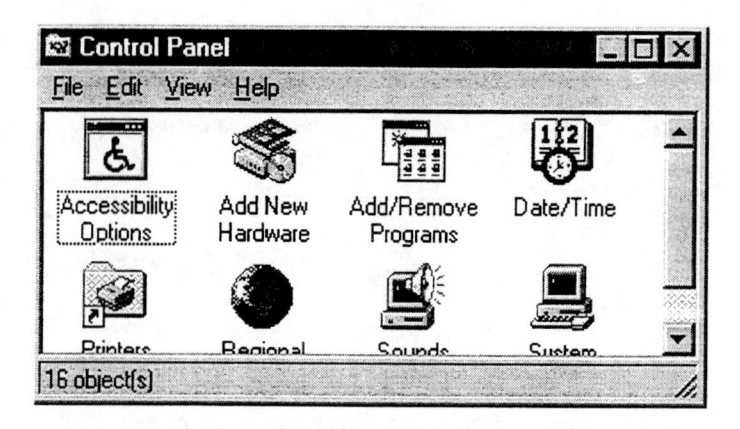

Figure 16-9 Control Panel Window—Select Add/Remove Programs

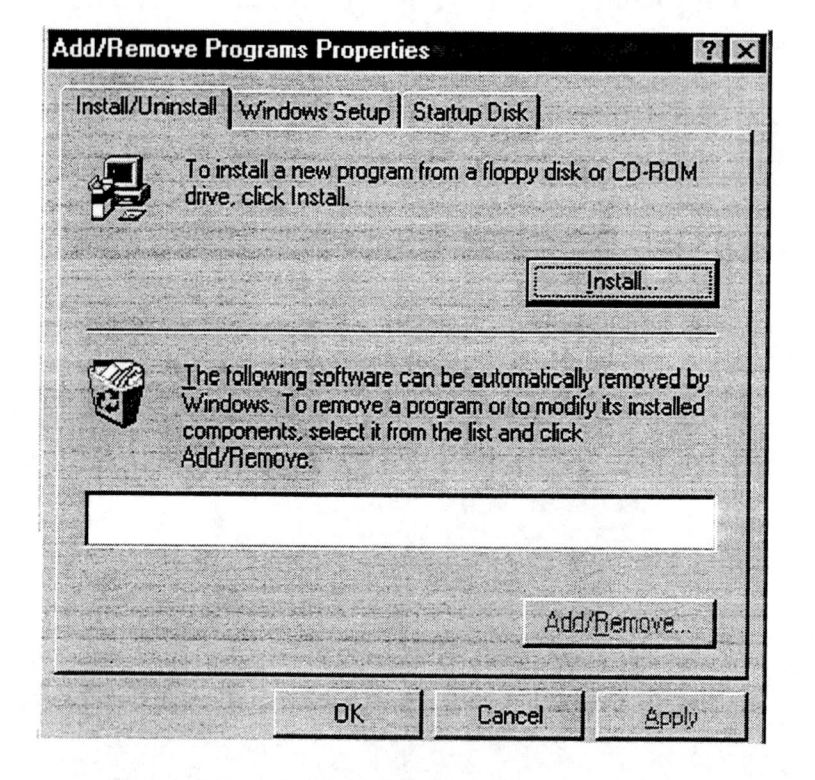

Figure 16-10 Add/Remove Programs Properties Window

Installing Hardware

Windows 95 supports the Plug and Play standard. If the hardware device you want to install (such as a modem, a Network Interface Card (NIC), or a sound card) is Plug and Play, then all you have to do is insert the card in any available expansion slot and turn on your computer. Windows 95 will install the hardware with a minimum of user interaction—sometimes none at all.

If the hardware device you want to install is not Plug and Play, you must install it manually. First put the device in your computer and start your computer. Once your computer has started, click the Start button, point to Settings, and click Control Panel. Double-click the Add New Hardware icon and follow the instructions on your screen.

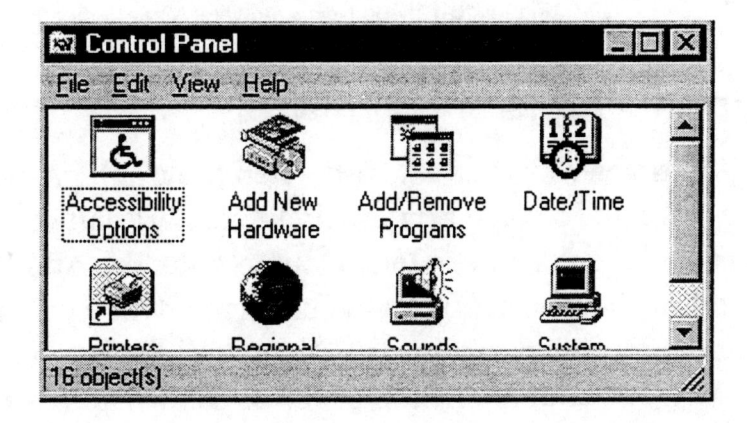

Figure 16-11 Control Panel Window—Select Add New Hardware

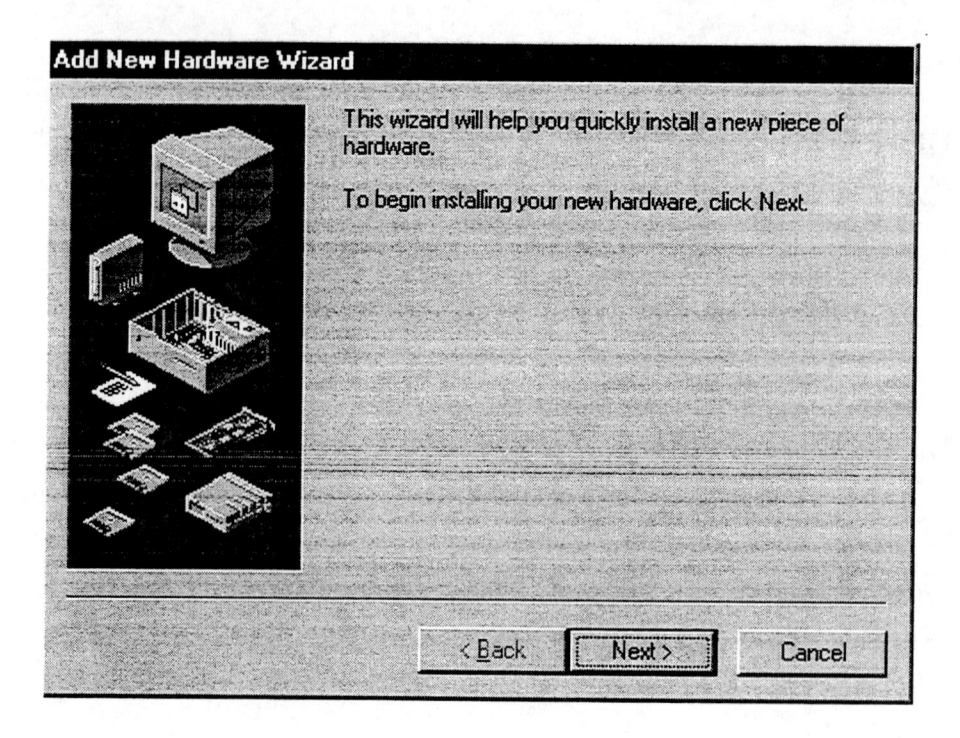

Figure 16-12 Add New Hardware Wizard Screen

Adding Printers

The Add Printer wizard is a step-by-step program that helps you set up a printer. To add a printer, click the Start button, point to Settings, and then click Printers. Double-click the Add Printer icon. Follow the instructions on your screen. Once you have installed your printer, that printer's icon will appear in the Printers folder. You can double-click this icon to see what documents are printing or waiting to be printed. You can also manage printing from this icon.

Chapter 16: Windows 95

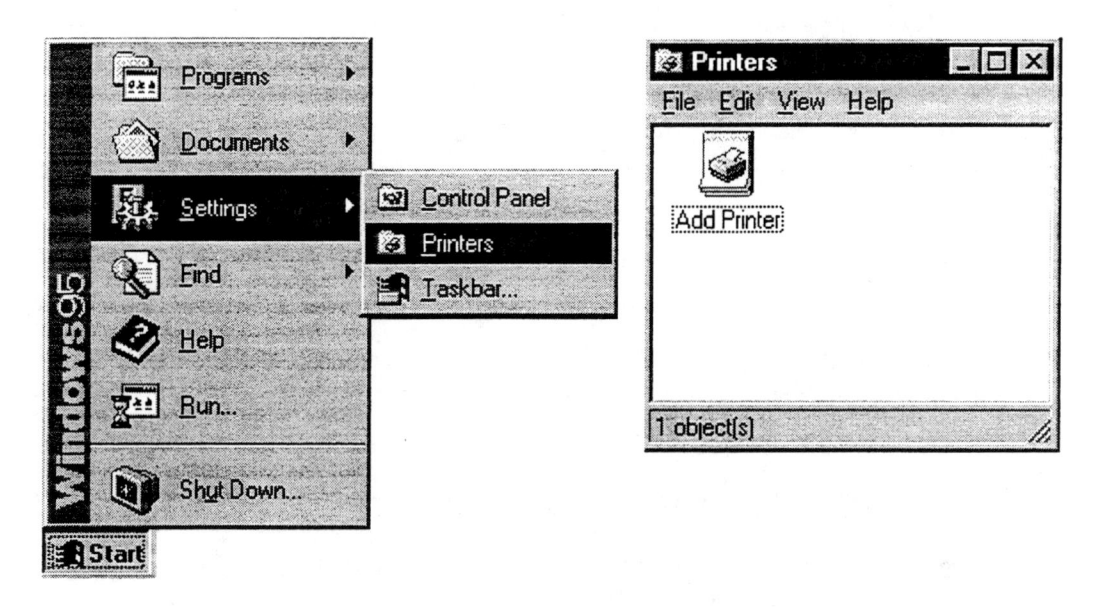

Figure 16-13 Accessing the Add Printer Screen

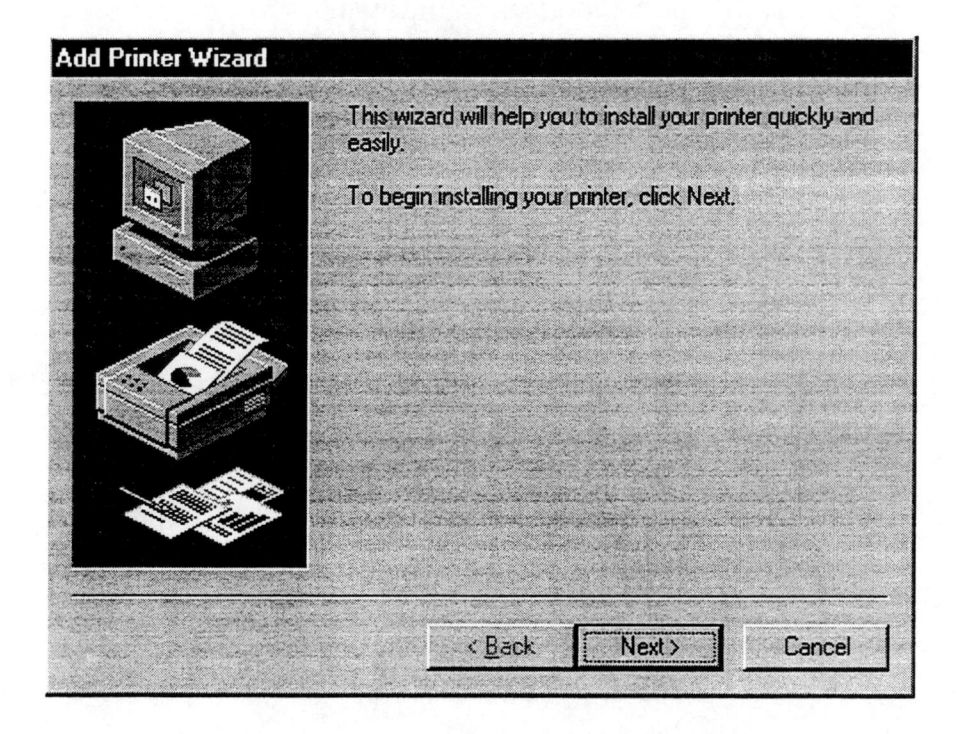

Figure 16-14 Add Printer Wizard Screen

Creating Shortcuts

Shortcuts provide easy access to programs and documents you use often. Once you create a shortcut for a program or a document, you can run that program or open up that document just by double-clicking on that shortcut icon. You can create a shortcut to any object, including folders, disk drives, other computers (if connected to a network), and printers. There are three ways to create a shortcut: Windows Explorer, the Find Files or Folders command, and the right mouse button.

Creating Shortcuts Using Windows Explorer

Open up Windows Explorer (make sure it does not fill the whole screen) and locate the item for which you want to create a shortcut. Using the left mouse button, drag the item to the desktop. Then release the mouse button. Windows 95 will create a shortcut for that item.

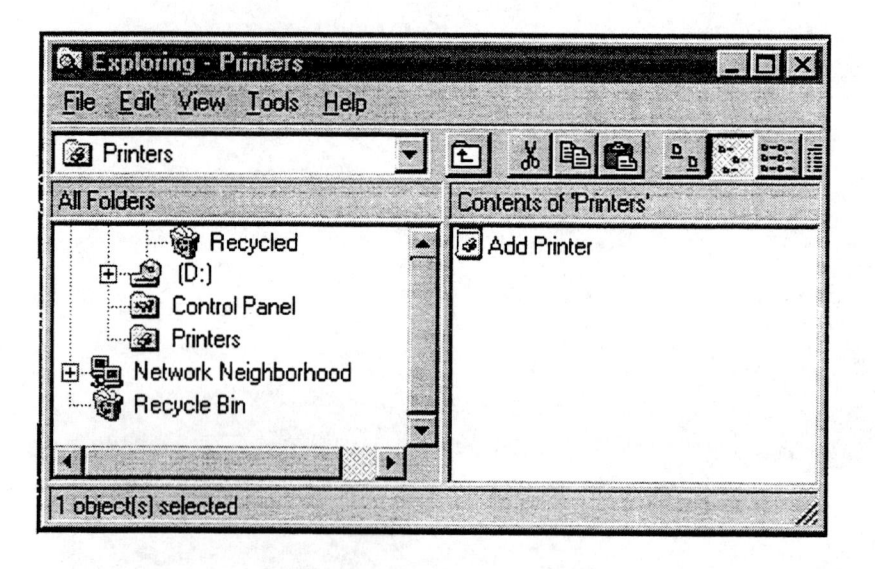

Figure 16-15 Using Windows Explorer to Create a Shortcut

Chapter 16: Windows 95

Creating Shortcuts Using the Files or Folders Command

Click the Start button, move the pointer to Find, then click Files or Folders. Enter the name of the item for which you want to create a shortcut in the Named box, then click the Find Now button. Windows will display the results of its search. Locate the desired item in the results. Using the left mouse button, drag the item to the desktop. Then release the mouse button. Windows 95 will create a shortcut for that item.

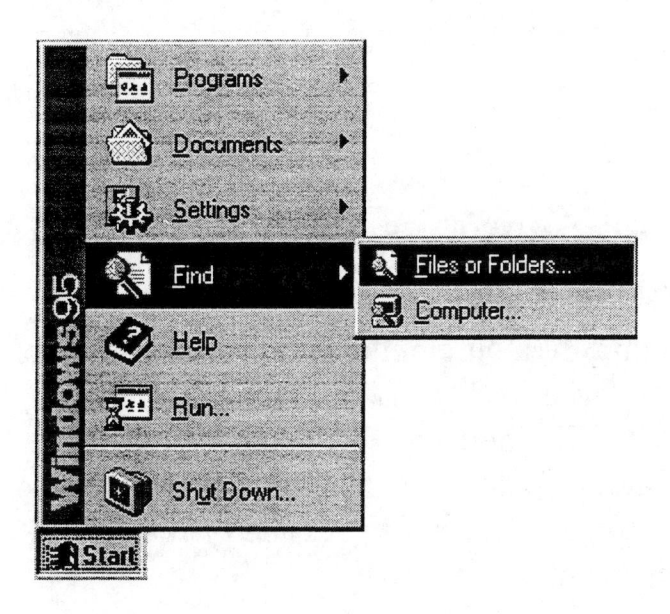

Figure 16-16 Accessing the Files or Folders Command

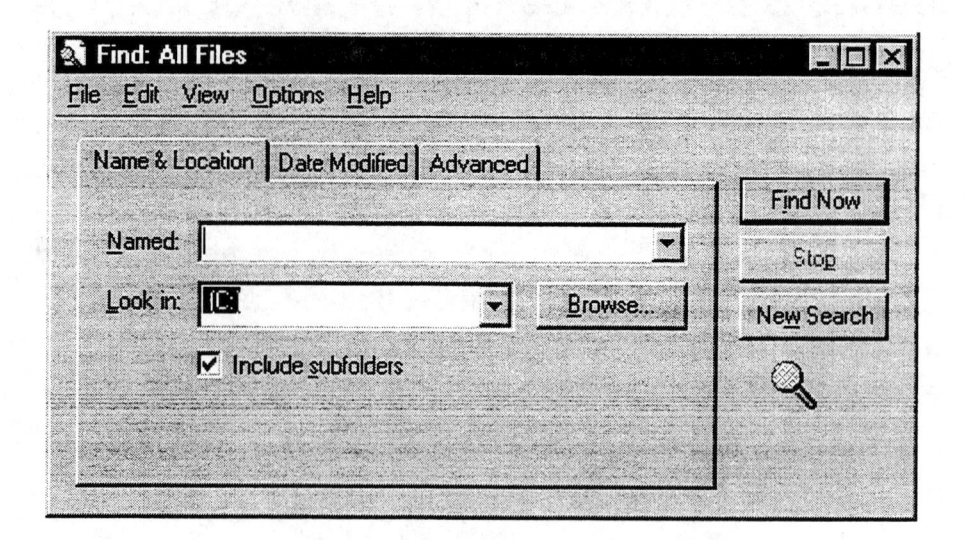

Figure 16-17 Files or Folders Command Window

Creating Shortcuts Using the Right Mouse Button

Click your right mouse button in any blank space on the desktop (that is, anywhere other than on an icon) and move the pointer to New, then click Shortcut. Now you can either type the name of the item (including path and extension, if required) or click the Browse button to select the path and/or filename. Then follow the instructions on your screen.

Setting Up a Network

A network is defined as a group of computers connected together in such a way that they can share information (data) and resources (printers, modems, etc.). There are four networking components included in Windows 95:

- Client software
- Network adapter
- Protocol
- Services

Client Software

This software enables you to connect to network resources, such as folders and printers, that have been shared on different computers on the network. With Client for NetWare Networks, you can connect to Novell NetWare servers. With Client for Microsoft Networks, you can use resources shared on computers running Microsoft Windows 95, Windows for Workgroups, Windows NT, and other Microsoft operating systems.

Network Adapter

A network adapter is an expansion card (commonly known as a Network Interface Card, or NIC) that physically connects your computer to the network.

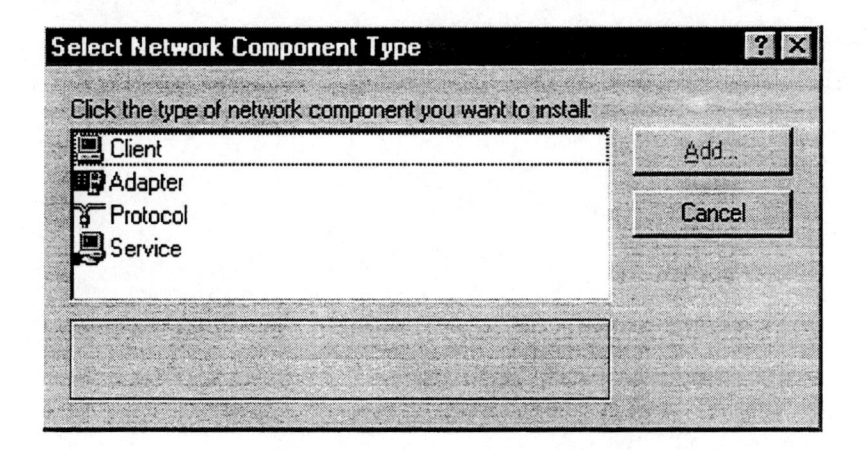

Figure 16-18 Selecting a Network Adapter Card

Protocol

A protocol is simply the language or set of languages that a computer uses to communicate. To be able to communicate with each other, two computers must use the same protocol. Several suites of protocols—such as TCP/IP, NetBEUI, IPX/SPX, etc.—are available with Windows 95.

Services

Network services are provided by networking hardware and software, and include such functions as file and printer sharing.

In order to set up a network, you must make sure that a NIC is present in your computer, properly installed (configured through Windows 95), and connected to your network. Click the Start button, point to Settings, and click Control Panel. Now double-click the Network icon. Click the Add button, click Adapter, and then click the Add button. Follow the instructions on your screen. If you are uncertain of what type of adapter you have, consult the documentation that came with it. When you set up a network adapter, Windows automatically sets up the other network components you need to use the network.

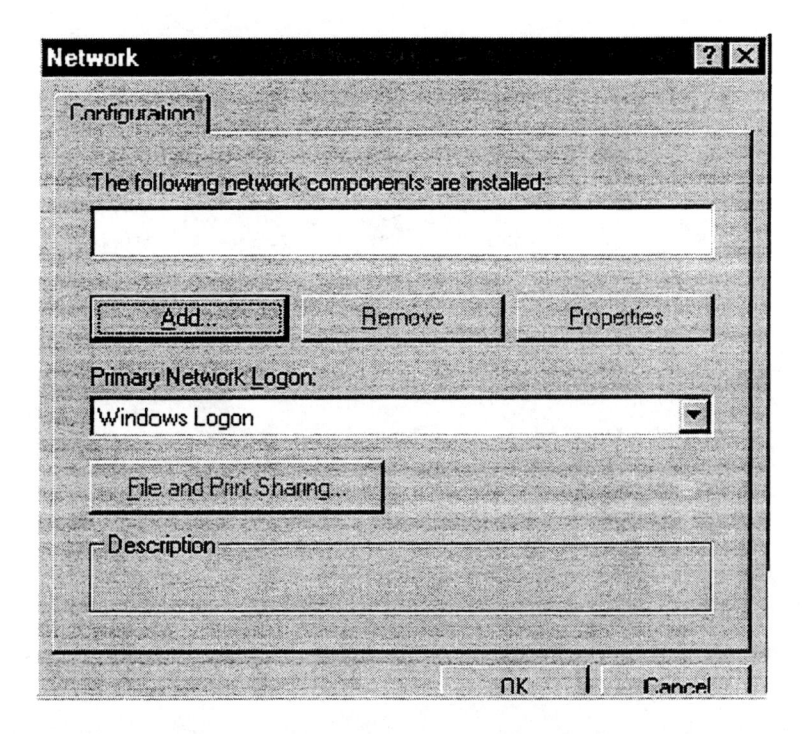

Figure 16-19 Network Window

After setting up the network software, you have to identify your computer to the network. Click on the Identification tab and type the requested information. This information is usually provided by your LAN administrator. When you have entered all the information, click OK.

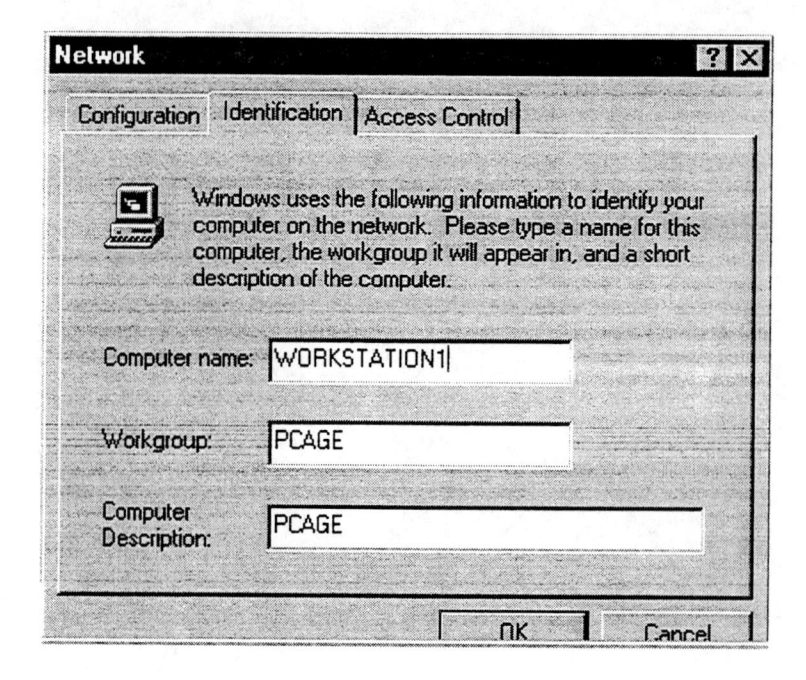

Figure 16-20 Identification Tab, Network Window

Sharing Folders and Printers

If you have installed file and printer servers for Microsoft networks, NetWare networks, or both, you can allow other computers on the network to share your files, your folders, and any printers attached to your computer.

Windows 95 provides both share-level access control and user-level access control. In share-level access control, you assign a password to each shared resource, and a user wanting to use the resource must know this password. In user-level access control, each user is assigned a password to log on to the network, and each shared resource is configured so that certain people or groups are authorized to use it.

Sharing a Folder in Share-Level Access Control

Locate the folder you want to share, click on it with your right mouse button, and click on Sharing. Click the options you want, then click OK.

Sharing a Printer in Share-Level Access Control

Click the Start button, point to Settings, and click Printers. From the Printers window, click the printer you want to share with your right mouse button, then click Sharing. Click the options you want, then click OK.

Maintaining and Optimizing Your System

Windows 95 provides the following maintenance tools:

- ScanDisk
- Disk Defragmenter
- DriveSpace

If used regularly, these tools can help maintain your hard disk. They can be found in the System Tools folder, which is in the Accessories folder.

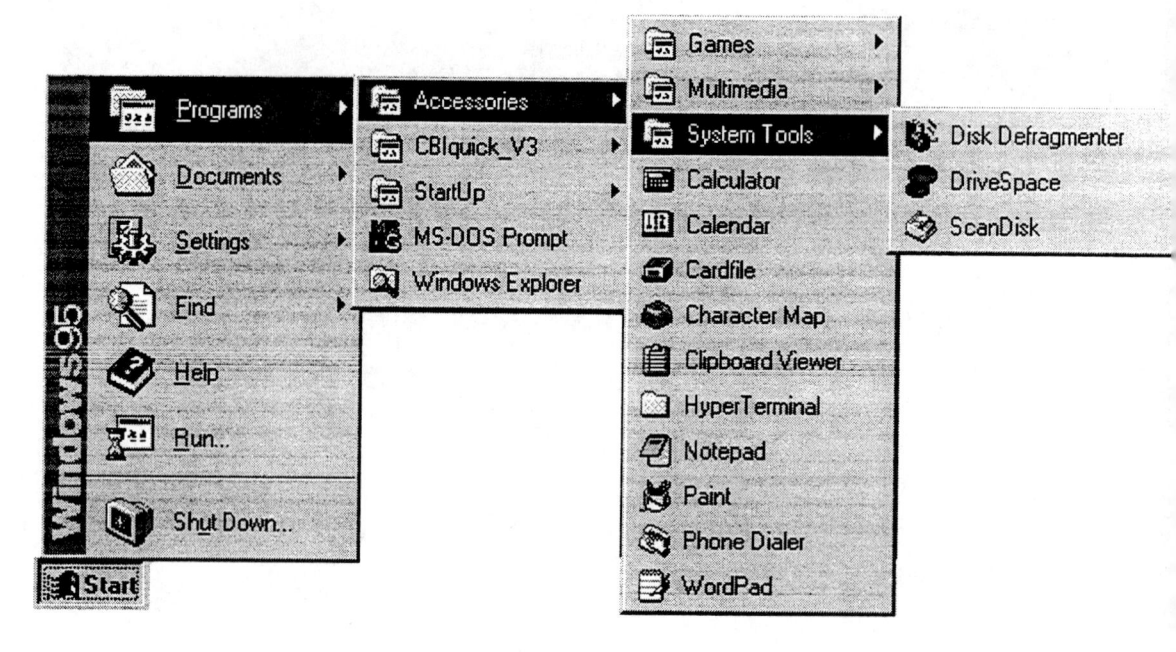

Figure 16-21 System Tools Folder

ScanDisk

ScanDisk can check for data error in the files and folders on your hard disk. It can also check the physical surface of your hard disk for damage.

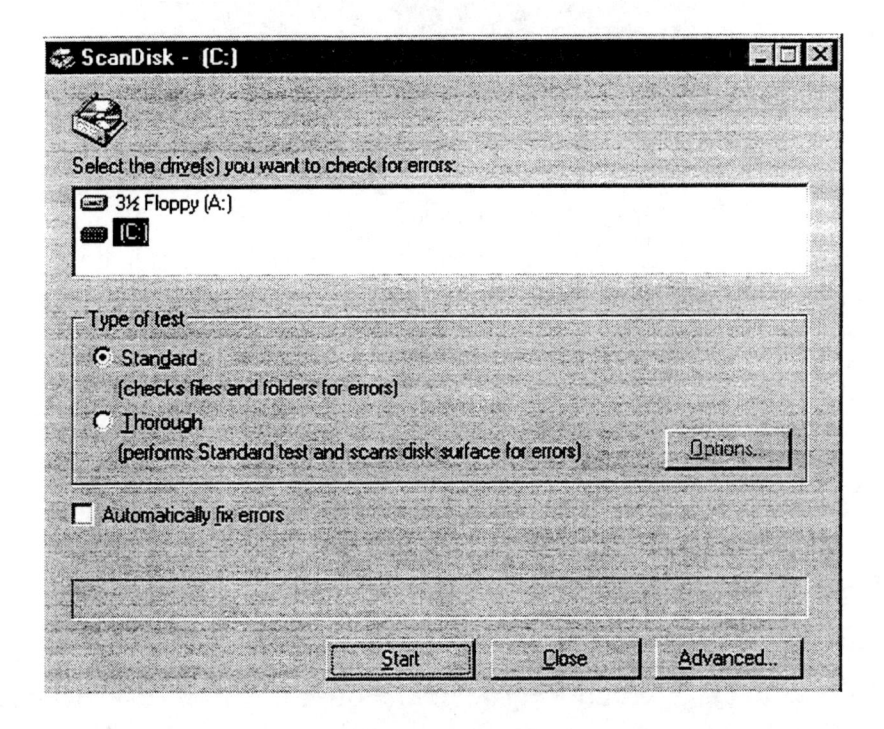

Figure 16-22 ScanDisk Window

Disk Defragmenter

Over long periods of time, as you save files on your hard disk, your files can become divided into fragments that are stored in different locations on the disk. The disadvantage of this condition (known as fragmentation) is that it makes I/O response slower. Windows 95 provides you with a tool called Disk Defragmenter to take care of this problem.

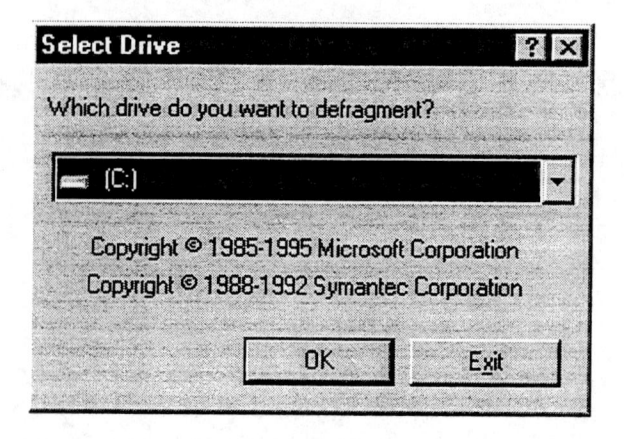

Figure 16-23 Disk Defragmenter Window

DriveSpace

Windows 95 also supports disk compression. By using DriveSpace to compress data, you can free up space on your hard disk or on a floppy disk. Using this utility on an uncompressed drive can increase your free space by as much as 100 percent.

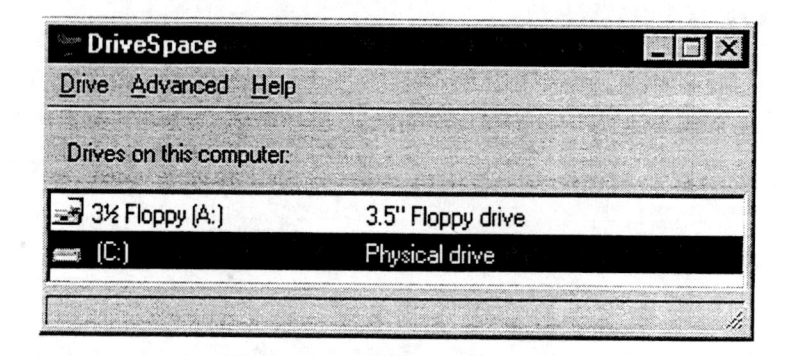

Figure 16-24 DriveSpace Window

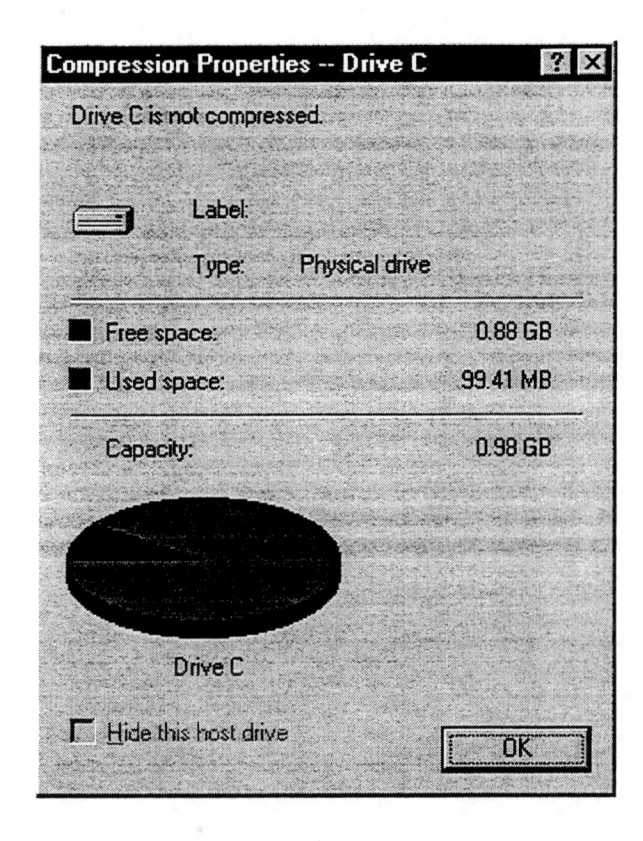

Figure 16-25 Compression Properties Window

Appendix A   Microprocessor Comparison Tables

Please refer to the following key to interpret the information in the tables below. The specifications listed in the key are only a small sampling of CPU specifications.

CPU Specification	Explanation	Unit of Measure
ExCF	The external clock rate of the CPU. This is the rate at which data is transported across the data bus.	MHz (megaHertz)
InCF	The internal clock frequency of the CPU. Generally, an operation is executed at each clock transition.	MHz (megaHertz)
WS	Word size. This is the largest piece of data that can be operated on in one operation.	Bits
DP	Data path. This is the largest piece of data that can be transported into the chip in one operation.	Bits
#IP	Number of instruction pipelines. This is the number of processes that the chip can run simultaneously.	Units
Mem	Memory. This is the maximum amount of memory the chip can use.	MB (megabytes)
MC	Math coprocessor. This allows for "Floating-Point" calculations, which in turn allows for better performance of certain graphics and spreadsheet applications.	Not applicable
InC	Internal cache. This is the amount of internal high-speed memory the CPU possesses.	k (kilobits— *not* kilobytes)

© 1998 · PC Age, Inc. All Rights Reserved · 20 Audrey Place · Fairfield, NJ 07004 · U.S.A. · Tel: 973-882-5002
www.pcage.com

Intel® Processors

CPU Type	ExCF	InCF	WS	DP	#IP	Mem	MC	InC	Note
8088	8	8	16	8	1	1	No	0	Equal to 29,000 transistors
8086	8	8	16	16	1	1	No	0	
80186	16	16	16	16	1	1	No	0	Not popular.
80286	20	20	16	16	1	16	No	0	Used in IBM AT.
80386DX	40	40	32	32	1	4096	No	0	Equal to 250,000 transistors.
80386SX	25	25	32	16	1	16	No	0	Note difference (shaded)
80486SLC	25, 33	25, 33	32	32	1	64	No	8	Low power drain. Popular in laptops.
80486DX	25, 33, 50	25, 33, 50	32	32	1	4096	Yes	8	Equal to 1.25 million transistors.
80486SX	20, 25, 33	20, 25, 33	32	32	1	4096	No	8	Identical to 80486DX, except math coprocessor is disabled.
80486DX2	20, 25, 33	40, 50, 66	32	32	1	4096	Yes	8	Referred to as "overdrive chip". Operates internally at twice the clock rate (see shaded area).
80486DX4	25, 33	75, 100	32	32	1	4096	Yes	8	Includes power management on chip; uses clock tripler.

ExCF	=	External Clock Frequency	InCF	=	Internal Clock Frequency
WS	=	Word Size	DP	=	Data Path
#IP	=	# of Instruction Pipelines	Mem	=	Memory
MC	=	Math Coprocessor	InC	=	Internal Cache

Intel® Processors

(continued)

CPU Type	ExCF	InCF	WS	DP	#IP	Mem	MC	InC	Note
Pentium	60, 66	60, 66, 75, 90, 100, 125, 133, 150, 166	32	32	2	4096	Yes	16	Equal to 3.1 million transistors. Best buy: InCF speed greater than 75 MHz runs cooler.
Pentium Pro	60, 66	166, 200, 233, 300	32	64	3	4096	Yes	16	Equal to 20 million transistors. Good for full 32-bit operating systems; not recommended for 16-bit operating systems.
Pentium MMX	60, 66	133, 150, 166, 200, 233	32	64	2	4096	Yes	16	New processor instruction codes; improved multimedia and communications performance.
Pentium II	60, 66	166, 200, 233, 300	32	64	3	4096	Yes	16	Can be considered a Pentium Pro with better 16-bit support and MMX.

ExCF = External Clock Frequency
InCF = Internal Clock Frequency
WS = Word Size
DP = Data Path
#IP = # of Instruction Pipelines
Mem = Memory
MC = Math Coprocessor
InC = Internal Cache

Note: MMX is not an acronym and does not officially stand for anything, but it is often referred to as "multimedia extension".

Cyrix® Processors

CPU Type	ExCF	InCF	WS	DP	#IP	Mem	MC	InC	Intel Equivalent
80486SLC	25,33	25,33	32	32	1	16	No	1	386SX
80486SLC	25	50	32	32	1	16	No	1	386SX
80486DLC	33	33	32	32	1	4096	No	1	386DX
80486DX	33, 40, 50	33, 40, 50	32	32	1	4096	Yes	8	486DX
80486DX2	25, 33, 40	50, 66, 80	32	32	1	4096	Yes	8	486DX2
586		100, 120	32	64	1	4096	Yes	16	Pentium
6x86	50, 55, 60, 66, 75	100, 110, 120, 133, 150	32	64	2	4096	Yes	16	Pentium

ExCF = External Clock Frequency
InCF = Internal Clock Frequency
WS = Word Size
DP = Data Path
#IP = # of Instruction Pipelines
Mem = Memory
MC = Math Coprocessor
InC = Internal Cache

AMD® Processors

CPU Type	ExCF	InCF	WS	DP	#IP	Mem	MC	InC	Intel Equivalent
386SE	25,33	25,33	32	16	1	16	No	0	386SX
386DE	33, 40	33, 40	32	32	1	4096	No	0	386DX
486DXLV	33	33	32	32	1	4096	Yes	8	486DX
486SXLV	33	33	32	32	1	4096	No	8	486SX
486DX	33, 40	33, 40	32	32	1	4096	Yes	8	486DX
486SX	33, 40	33, 40	32	64	1	4096	No	8	486SX
486DX2	25, 33	50, 66	32	32	1	4096	Yes	8	486DX2
486DX2-80	40	80	32	32	1	4096	Yes	8	486DX2
486SX2-50	25	50	32	32	1	4096	Yes	8	486SX
Am5x86	33	133	32	32	1	4096	Yes	16	Pentium 75

Appendix B   DOS Error Messages

ABORT, RETRY, IGNORE, FAIL?

DOS failed to recognize an instruction it was given. This problem may also be caused because of a disk error or a device error. You may choose one of the four options (it is not recommended to use the last two options).

ACCESS DENIED

You have attempted to open a file that is write-protected or labeled read-only. You may also get this error message if you use the TYPE command on a subdirectory, or the CD or CHDIR command on a file. Make sure your disk is not write-protected. Use the ATTRIB command to change the attributes if necessary.

ALL FILES IN DIRECTORY WILL BE DELETED! ARE YOU SURE?

You are about to delete all files in the specified directory or the currently logged directory. You can choose either of the two options.

ANSI.SYS MUST BE INSTALLED

You did not install ANSI.SYS, or you have not used the correct syntax in the CONFIG.SYS file. Edit the ANSI.SYS command and reboot your system.

ATTEMPT TO REMOVE CURRENT DIRECTORY

You used the RD or RMDIR command using the name of your currently logged directory (you cannot remove the root directory). Use CD to log onto the parent directory, then try again.

BAD COMMAND OR FILE NAME

DOS did not recognize your command. Most of the time this problem is caused by a spelling error. Make sure that you have entered the command correctly. Also, make sure that the command file is in either the specified directory or the search path.

BAD OR MISSING COMMAND INTERPRETER

This problem relates to COMMAND.COM. Most of the time it means that your COMMAND.COM file is corrupted. Reboot your system with a bootable floppy disk. Make sure that the version of COMMAND.COM matches the version of your system before replacing your COMMAND.COM file.

BAD OR MISSING DRIVER

DOS cannot locate the device driver file, or the file has become corrupted. Check the syntax in CONFIG.SYS first. If it is correct, then copy a new device driver in the specified directory and reboot your system.

BAD OR MISSING KEYBOARD DEFINITION FILE

Either DOS cannot locate your KEYBOARD.SYS file, or the file has become corrupted. KEYBOARD.SYS and KEYB.EXE are both located in the same directory. You may have to copy a new file from the backup.

Appendix B: DOS Error Messages

BAD UMB NUMBER

If you receive this message, try to fix it by running MEMMAKER. If you receive this message when using the LH or LOADHIGH command, remove the /L switch and any accompanying parameters. If the problem still persists, get your hardware checked by a professional.

BATCH FILE MISSING

This message appears when a batch file has erased itself. First try to recover it using the UNDELETE command. If this doesn't work, make another batch file.

BOOT ERROR

If this message appears, check your computer's setup parameters using your system's setup utility. It is recommended that you get a professional to solve this problem.

CANNOT CHDIR TO PATH

CHKDSK cannot verify the existence of a subdirectory reported in the FAT. Run CHKDSK with the /F option to correct this problem.

CANNOT CHDIR TO ROOT

CHKDSK cannot locate the start of the root directory. Reboot your computer and re-invoke the command. If the problem persists, you will have to format your drive.

CANNOT DO BINARY READS FROM A DEVICE

You have just used the COPY command with the /B switch. Try the same command with the /A switch.

CANNOT FIND SYSTEM FILES

You get this message when you try to load the operating system from a drive that does not contain the system files. Use the SYS command to transfer the system files and reboot your system. Your CONFIG.SYS file and your AUTOEXEC.BAT file should be in the root directory. If the problem persists, you may have to format your drive using the FORMAT /S command.

CANNOT LOAD COMMAND, SYSTEM HALTED

This problem is caused by COMMAND.COM, as DOS is unable to reload the command processor. If you are using the COMSPEC environment variable, make sure the specified path is right.

CANNOT OPEN SPECIFIED COUNTRY INFORMATION FILE

You have used an invalid file name for the country code page in the CONFIG.SYS file. Check the syntax of the COUNTRY command in CONFIG.SYS. If the problem remains, then check the syntax of the KEYB, NLSFUNC, and MODE commands in the AUTOEXEC.BAT file.

CANNOT PERFORM A CYCLIC COPY

You have probably tried to copy files onto themselves, or have used the XCOPY command with the /S switch in nested subdirectories. Correct the syntax and try the command again.

CANNOT READ FILE ALLOCATION TABLE

The file allocation table has become corrupted. This problem can usually be fixed by using the CHKDSK command. If not, get it checked by a professional.

CANNOT SETUP EXPANDED MEMORY

EMM386 cannot initialize properly. Check its syntax in the CONFIG.SYS file. Also, make sure that HIMEM.SYS is loaded before this command. If the problem remains, get it checked by a professional.

CURRENT DRIVE IS NO LONGER VALID

Your currently logged drive has no disk in it, the drive door is open, or the drive is unrecognizable on a network. Check the cause that relates to your particular problem.

DATA ERROR

DOS has detected inconsistencies in data while reading or writing a file. If prompted to Abort or Retry, press R a few times. If the operation does not continue, press A, then run the CHKDSK command. You may have to format your disk.

DIRECTORY ALREADY EXISTS

You tried to create a directory with the same name as a directory that already exists at that level. Try the same command with a different directory name.

DISKETTE BAD OR INCOMPATIBLE

The disk you are using has an incorrect format. It may be copy-protected, or may contain data error. If the disk is not copy-protected, use the CHKDSK command to fix this problem.

DRIVE OR DISKETTE TYPES NOT COMPATIBLE

This message appears when you try to use the DISKCOMP or DISKCOPY command on drives with two different format types. Try using the FC or XCOPY command.

DUPLICATE FILE NAME

You have tried to use a file name that already exists. Use a different name.

DUPLICATE REDIRECTION

You have used the redirection symbols to read data from a file being written. Revise the syntax, using a unique output file name.

ERROR IN EXE FILE

The application you are trying to run has a corrupted executable file. It may also be incompatible with your current DOS version. First check the DOS version. If it is correct, then copy a new executable file from the backup.

ERROR LOADING OPERATING SYSTEM

Refer to "CANNOT FIND SYSTEM FILES".

ERROR READING DIRECTORY

The file allocation table or subdirectory structure has become corrupted. Reformat your disk. If the problem remains, get your system checked by a professional.

ERROR READING (OR WRITING) DRIVE

This message usually appears when you are trying to use a corrupted disk. Try another disk. If the problem remains, try to reboot your system. If you cannot fix this problem, have your drive serviced.

ERROR READING PARTITION TABLE

The hard disk's partition is unusable. Use FDISK to set up the partition table before attempting to format the disk. If the problem remains, check with a professional.

ERROR READING SYSTEM FILES

Refer to "CANNOT FIND SYSTEM FILES".

EXEC FAILURE

Refer to "ERROR IN EXE FILE".

EXPANDED MEMORY MANAGER NOT PRESENT

You forgot to install the expanded memory manager before installing the drivers that require this memory. Make sure that it is present in the CONFIG.SYS file. Also, check the order of execution.

EXTENDED MEMORY MANAGER NOT PRESENT

You must install XMS extended memory HIMEM.SYS before installing drivers that require this memory, such as EMM386.EXE. Also, check the order of execution.

FCB UNAVAILABLE

DOS has been instructed to access a file control block that is out of range. Revise the FCBS command in CONFIG.SYS. If the problem remains, seek help from a professional.

FILE ALLOCATION TABLE BAD

The file allocation table has become corrupted. Run CHKDSK. You may have to format the disk. If the problem remains, get it checked by a professional.

FILE CANNOT BE COPIED ONTO ITSELF

You have specified the same file as both the source and the target. This problem usually arises when the wildcard characters are not used properly. Change the specification for the source or the target.

FILE CREATION ERROR

There are several things that can cause this problem: there is not enough space on the disk or the chosen subdirectory; the file you tried to create already exists and is read-only; you tried to rename a file using a file name that already exists; or you attempted to redirect output to an invalid file name. If you get this message while creating a file in the root, make sure the total number of files in the root does not exceed 512. Delete some files to make space, if necessary. Also, use the ATTRIB command if your problem is caused by the file attributes.

FILE IS CROSS-LINKED

CHKDSK has found two files that share the same area of the disk. Use the /F switch with the CHKDSK command to take care of this problem.

FILE NOT FOUND

The requested file was not found on the currently logged directory. This message also appears if the subdirectory is empty or if there is no path specified in the PATH command.

GENERAL FAILURE

The disk in the drive is not formatted or is formatted for another system. Reformat the disk. If the problem persists, check with a professional.

I/O ERROR

DOS has discovered a RAM error while processing. Check the status of all memory-resident applications for conflicts with the current application. If the problem persists, get it checked by a professional.

INCOMPATIBLE SWITCHES

You have used mutually exclusive options on the command line. Review the syntax and enter the correct options.

INCORRECT DOS VERSION

You entered a DOS external command for a version of DOS other than the one you are currently using. Reboot your system with the correct version of DOS, or use the correct executable file for the command.

INSUFFICIENT DISK SPACE

You have used all of your disk space. Run CHKDSK to reclaim some space that may be occupied by lost clusters. You may also have to delete some files.

INSUFFICIENT MEMORY

You don't have enough RAM to process the command. Try removing some memory-resident files. You may have to reboot your system. In the long run, you will need to add more RAM to your system.

INTERNAL ERROR

This message appears because of memory conflict or another technical error. It is advisable to get it checked by a professional if you see the same message again after rebooting your system.

INVALID COMMAND.COM

Refer to "BAD OR MISSING COMMAND INTERPRETER".

INVALID DATE

DOS cannot recognize the date format you have entered, or you have entered a nonexistent date. Check your entry and try again.

INVALID DEVICE REQUEST

DOS was unable to process a device driver command. Check the configuration syntax for the driver file and try again. If the syntax is correct, recopy the driver file and try again. If the problem persists, it may mean that you have a hardware failure. Seek help from a professional.

INVALID DIRECTORY

You have entered an invalid directory name or the name of a directory that does not exist. Sometimes DOS also discovers an invalid directory. Check the syntax and try the command again. If DOS discovered the invalid directory, you have to reformat the disk.

INVALID DISK CHANGE

DOS has discovered that you have changed disks during processing. Put the original disk back. If this does not solve the problem, then it is a hardware problem and should be fixed by a professional.

INVALID DRIVE IN SEARCH PATH

You have referenced a drive letter in the PATH command that is out of range on your system. Edit the PATH command syntax or the LASTDRIVE command in CONFIG.SYS.

INVALID DRIVE SPECIFICATION

You have entered the letter of a drive that does not exist. Enter a different drive letter, or assign the drive letter using the ASSIGN or SUBST command.

INVALID FILENAME

You have entered a file name containing invalid or wildcard characters, or you have used a reserved device name as a file name. Try the command again using a different file specification.

INVALID FUNCTION PARAMETER

Refer to "INVALID PARAMETER".

INVALID KEYWORD

Refer to "INVALID PARAMETER".

INVALID MEMORY BLOCK ADDRESS

Refer to "BAD UMB NUMBER".

INVALID NUMBER OF PARAMETERS

Refer to INVALID PARAMETER.

INVALID PARAMETER

You have not specified the correct option switches on the command line, you have duplicated parameters, or you have combined parameters illegally. Review the command syntax and try the command again.

INVALID PARAMETER COMBINATION

Refer to INVALID PARAMETER.

INVALID PARTITION TABLE

DOS has detected an error in the fixed disk's partition information. Run FDISK to initialize a valid partition table.

INVALID PATH

You have used the RD or RMDIR command using the name of the currently logged directory. Log onto the parent directory and try the command again. This problem also arises when you specify a nonexistent directory, or one that DOS cannot find. Check the drive and path specification, the spelling of the directory name, and the settings of the PATH and APPEND commands.

INVALID PATH, NOT DIRECTORY, OR DIRECTORY NOT EMPTY

DOS is not able to locate the specified directory, you have entered a file name in place of a directory name, or the directory contains files (or other nested subdirectories) and cannot be removed. Check the spelling of the directory name, or list the contents of the directory. If it appears empty, it may contain hidden files. Use the DIR /A:H command to reveal any possible hidden files.

INVALID SWITCH

Refer to INVALID PARAMETER.

INVALID SYNTAX

DOS could not process the syntax you entered. Review the command syntax and try again.

INVALID TIME

DOS cannot recognize the time format you have entered. Check your entry and try again.

MEMORY ALLOCATION ERROR

DOS was not able to configure RAM properly. Reboot the computer. If the problem persists, reboot your system again using a bootable floppy disk and use the SYS command to copy new system files. If the problem still remains, get it checked by a professional.

MUST SPECIFY ON OR OFF

You have entered an invalid parameter for a command that requires a parameter of ON or OFF. Review the syntax and try the command again.

NO EXTENDED MEMORY AVAILABLE

The XMS extended memory has been allocated to other applications and resident functions. Deactivate other drivers to make room for your RAM disk.

NO FIXED DISKS PRESENT

DOS was not able to detect the presence of the fixed disk drive. Check your computer's setup parameters for the correct drive type. If you cannot solve the problem yourself, seek the help of a professional.

NO ROOM IN DIRECTORY

You have exceeded the limit on the number of allowable files in the root directory (512 files). Copy the file to a subdirectory or a different disk.

NON-SYSTEM DISK OR DISK ERROR

DOS cannot find system files on the current disk. Insert a disk containing system files, or boot from the hard disk if it contains system files.

NOT ENOUGH MEMORY

Refer to "INSUFFICIENT MEMORY".

OUT OF ENVIRONMENT SPACE

You have initialized too many operating system variables. Remove some variables using the SET command or increase the size of the environment space using the SHELL command in the CONFIG.SYS file, then reboot your computer.

PARAMETER FORMAT NOT CORRECT

Refer to INVALID PARAMETER.

PARAMETER VALUE NOT ALLOWED

Refer to INVALID PARAMETER.

PARAMETERS NOT COMPATIBLE

Refer to INVALID PARAMETER.

PARSE ERROR

You have entered a syntax error, but DOS cannot locate COMMAND.COM to display a more precise error message. Reboot the computer if necessary, check the syntax of the command, and try using it again.

PATH NOT FOUND

Refer to "INVALID PATH".

PRINTER ERROR

DOS cannot send data to your printing device. Make sure the printer is attached, is turned on, and has paper. Also, make sure that the output has not been redirected to a different port.

READ FAULT ERROR

DOS cannot read data on the disk. Re-insert the disk in the drive and press R. If you still have problems, run CHKDSK. You may have to format the disk.

REQUIRED FONT NOT LOADED

DISPLAY.SYS has not been initialized to include the desired font. Edit your CONFIG.SYS file, increasing the number of subfonts, and reboot the computer.

REQUIRED PARAMETER MISSING

Refer to INVALID PARAMETER.

SAME PARAMETER ENTERED TWICE

Refer to INVALID PARAMETER.

SECTOR NOT FOUND

DOS has discovered a formatting error on the disk. Run CHKDSK. If this does not solve the problem, reformat the disk.

SEEK ERROR

Refer to "READ FAULT ERROR".

SHARING VIOLATION

You have attempted to reopen or write to a file that is already open. This usually happens on a network, but can also happen on single-user systems with SHARE installed. Use your current application's commands to close the data file before attempting to reopen it or write new data to it.

SOURCE DOES NOT CONTAIN BACKUP FILES

DOS was unable to locate the CONTROL.001 and BACKUP.001 files on the floppy disk. Check the disk and try again.

SPECIFIED COMMAND SEARCH DIRECTORY BAD

The SHELL command in the CONFIG.SYS file contains invalid information. Edit this line and reboot the system.

SPECIFIED DRIVE DOES NOT EXIST, OR IS NON-REMOVABLE

You have invoked a command intended for use on floppy drives, referencing a hard disk drive letter or drive letter that does not exist. Correct the syntax and try again.

SYNTAX ERROR

Refer to INVALID PARAMETER.

TOO MANY FILES OPEN

Refer to "TOO MANY OPEN FILES".

TOO MANY OPEN FILES

You have exceeded the maximum number of files that are allowed to be open on your system. Edit the FILES command in CONFIG.SYS and reboot the system.

TOO MANY PARAMETERS

Refer to INVALID PARAMETER.

TOO MANY REDIRECTIONS

You have redirected data output to a device that does not exist, or you have attempted to redirect data that has already been redirected. Correct the command line syntax for the device, or use a single redirection.

TRACK 0 BAD

DOS has detected disk errors in a critical portion of the disk. Reboot the system and try accessing the disk again. If this doesn't solve the problem, discard the disk. If the problem persists with a different disk, check with a professional.

UNABLE TO CREATE DIRECTORY

You have attempted to create a directory with the MD or MKDIR command, but either a directory of the same name already exists or you have reached the limit on the number of entries. Use a different name, or try to create the directory at another level. This problem may also be caused if your disk is write-protected.

UNRECOGNIZED COMMAND IN CONFIG.SYS

DOS could not recognize a command line in the CONFIG.SYS file when booting. Other messages that appear before this one may help you determine which line(s) is invalid. Edit your CONFIG.SYS file to correct this problem. You can also use the F8 key to check for the command that is creating the problem.

UNRECOGNIZED SWITCH

Refer to INVALID PARAMETER.

UNRECOVERABLE READ OR WRITE ERROR

DOS could not read data from or write data to the disk. The disk is probably damaged. Use a different disk to save the current data. Run CHKDSK to fix the damaged disk. You may even have to reformat the disk as a last resort.

WARNING! INVALID PARAMETER IGNORED

DOS cannot recognize a parameter you have entered on the HIMEM initialization line in CONFIG.SYS. Edit this line in the CONFIG.SYS file, then reboot your system.

WRITE PROTECT ERROR

DOS cannot write data to the disk because it is write-protected. Remove the write-protection tab and try the command again. If the error persists, use a different disk.